THE GREAT NUTRITION ROBBERY

THE GREAT
NUTRITION
ROBBERY »»»»»»»»»»»»

«««««««««« **Beatrice Trum Hunter**

CHARLES SCRIBNER'S SONS › NEW YORK

Copyright © 1978 Beatrice Trum Hunter

Library of Congress Cataloging in Publication Data

Hunter, Beatrice Trum.
    The great nutrition robbery.

    Bibliography: p. 199
    Includes index.
    1. Food, Artificial.   2. Food substitutes.
I. Title.
TX357.H83   362.5   78-1460
ISBN 0-684-15345-9

Portions of this book are based on articles
which appeared in *Consumer Bulletin* and in
Consumers' Research Magazine and are used
here with the kind permission of Consumers'
Research, Inc., Washington, New Jersey 07882.

1 3 5 7 9 11 13 15 17 19 V|C 20 18 16 14 12 10 8 6 4 2

PRINTED IN THE UNITED STATES OF AMERICA

To

Professor Ross Hume Hall,

with homage

CONTENTS

THE GREAT
NUTRITION
ROBBERY

1. The Great Pretenders

〈〈〈〈〈〈〈〈〈

The artificial synthesis of food is an example of the substitution of chemical manufacture for what had previously been a biological process. . . . It is well within the bounds of practicality to synthesize agreeable and attractive foodstuffs guaranteed to do to those who eat them no good at all.
—Dr. Magnus Pyke, *Synthetic Food* (London: John Murray, 1970)

Biologically, man appears to have changed little, if any, since he emerged as a distinct species about a million years ago. Although his powers of adaptation have taken him from the Himalayas to the depths of the sea and even into outer space, his internal environment is restricted to the narrowest biochemical and biophysical margins, which we can do little about.
—Raymond Shatin, M.D., quoted in *Medical World News*, 18 December 1964

In the near future we may all be drinking milk that didn't come from a cow or goat. It's all part of the explosion in fabricated foods, the sales of which the USDA estimates will total $23 billion in 1980, almost double the $13 billion in total sales for 1973. And by 1980, a Cambridge, Massachusetts, market analyst firm [Arthur D. Little, Inc.] predicts, the food industry will spend about $2.5 billion on substitute ingredients, compared to $1 billion in 1975. So it's goodbye, Elsie.
—*Food Management*, November 1975

The Great Nutrition Robbery

"Most People Have No Taste; It's Been Lost in the Process (Fresh Foods Taste Peculiar, If You Grew Up Eating Instant, Frozen or Canned)"
—headline, *Wall Street Journal,* 30 April 1974

Charles Grimm, director of flavor creation for International Flavor & Fragrances, Inc., was quoted saying that if he was developing a bread flavor today, he would aim at the taste of Wonder Bread rather than homemade bread.

"We've moved away from the utilization of fresh flavor. It isn't familiar anymore," says William Downey, chief flavor chemist for Fritzsche Dodge & Olcott, Inc. "Everybody likes to think fresh. But when designing a product, manufacturers want the taste people are accustomed to."

Another consultant commented: "There has been a very definite shift in food preference to the taste of processed foods. . . . If you gave someone ice cream made with fresh strawberries, you'd have a totally unacceptable product. People would say 'I wouldn't eat that artificial stuff.' "
—*Snack Food,* May 1976

The new president of Morton Foods criticized me for my exposé of his lemon cream pie which contains no real lemon, no cream and no eggs for that matter. He said that the Morton Company was doing society a real favor by putting artificial ingredients into their pies because this was getting us used to the future when all of our foods will be artificial.
—Marshall Efron, interviewed by Philip Nobile, "Uncommon Conversations." *Universal Press Syndicate,* 1972

In the best Horatio Alger tradition, yogurt has risen in 35 years from an obscure immigrant food from the wrong part of Europe into a spot on the menu of a "culture-conscious" U.S. consuming public. Also in the American tradition of making a good thing "even better"—it's now offered in such varieties as Swiss-style, soft-frozen, sundae-style, chocolate-covered on a stick, and as a "yogaburger." All the more amazing when one considers the unlikelihood of getting a germ-fearing, bug-killing, anti-bacterial society to accept yogurt at all. Or, some ask, is our currently marketed pasteurized, sterilized, preserved, sweetened product really yogurt?
—*Food Processing,* October 1976

The Great Pretenders

No one is quite sure what the American public is eating anymore.
—William C. Hale, senior project manager, food and agribusiness section,
Arthur D. Little, Inc., quoted in *Moneysworth,* 16 February 1976

If it's true, as they say, that we are what we eat, then what the hell are we?
—Nicholas von Hoffman, "Ersatz Food," Washington *Post,* 16 April 1971

>>>>>>>>>>

Natural versus Synthetic

Traditional foods, which have nourished human beings throughout the centuries, have undergone radical transformation. New technologies have been developed that make possible the use of substitute ingredients as partial or even complete replacers of traditional ones. With diminished or erratic supplies of traditional ingredients, which also have become more costly, replacement ingredients have become especially attractive to food processors. Replacers help keep manufacturing costs low and permit higher margins of profit. Although food quality is lowered, food prices are not. As will be demonstrated repeatedly, food products containing replacers retail at *higher* prices than traditional ones. Entirely new food products have been created which, although attractive and palatable, offer little or no nutritive value. Completely synthetic foods have been fabricated as substitutes for real ones. And this radical transformation of our foods continues, in ever more extreme fashion. This book explores in detail the reshaping of foods and raises questions about their impact on human health.

The Chemical Approach to Biological Processes

The philosophical basis of modern industrialized food production began in the nineteenth century. A passionate interest in scientific investigation led to a widespread belief that all substances could be analyzed, identified, classified, and synthesized. "Nature" could not

3

only be fully understood but could also be duplicated and even surpassed. That notion continues to flourish, as reflected in many advertising slogans. Hair dye is "better than natural"; egg substitutes are "so close to Nature—you'll never know!"; food colors are offered by a "technology that tops nature"; and food flavors are "taking nature further."

The notion that biological processes could be duplicated was basic to the work of a German chemist, Baron Justus von Liebig, who made an extensive study of "vegetable physiology" and synthesized urea. Ultimately, that achievement led to the synthesis and widespread use of chemical fertilizers. Since food itself is intimately linked with agricultural practices, Liebig's work represents a significant early manifestation of the synthesizing that now pervades all aspects of food production. Soil was no longer viewed as an organic entity but rather as something that could be put together in a mechanistic way by assembling parts. Holism gave way to fragmentation.

The industrial revolution contributed to the fragmentation of foods. Highly sophisticated machinery was developed for food processing: steel roller mills refined flour, other mills refined sugar.

Substances and techniques used in many unrelated industries found ready applications in food processing and little attention was given to either the hazards or the nutritional alterations that resulted. In 1856 Sir William Henry Perkins synthesized mauve dye. Soon hundreds of dyes were synthesized from coal tars for industrial uses and many of those dyes were used in the food supply, replacing plant dyes and mineral pigments. Similarly, many industrially used chemical additives were applied to food production, with little or no information about their safety. At that time, too, industrial machinery, such as spinning and extrusion equipment, was adapted for food processing.

Later, such machine processes were radically transformed by means of chemical processes. And finally, the use of substitute and synthetic ingredients on a large scale led to "the new look in food."

Our Synthetic Environment

We live in times in which synthetics and substitutes have replaced many natural substances in everyday life. We have grown accustomed to those replacers and in some instances find them more convenient or agreeable than the originals.

Our homes may be built of imitation stone or brick, roofed with fake slate (or in England with fake thatching), braced with simulated wood beams, and lighted with make-believe gas or kerosene lamps. Our dining table may be set with plastic tablecloths, dishes, and bowls. The little brown stoneware jug or the straw on the chianti wine bottle may be of molded plastic.

The decorative plastic scuttle may stand beside the fake fireplace with its simulated logs. The hearth may be swept with a broom made of plastic fibers which is supposed to resemble a corn broom. Artificial flowers or plants may rest on a "genuine reproduction" of an antique table. Draperies, upholstery, rugs, and other coverings, as well as our clothing, may be of synthetic fibers: simulated suede, imitation leather, fake fur, or materials advertised as "genuine sham."

The simulated marble dressing table may hold costume jewelry, including fake pearls, paste diamonds, or a synthetic star sapphire. Nearby there may be a supply of false eyelashes, false nails, false teeth, and a hairpiece.

As part of that increasingly synthetic environment, we have inevitably come to accept the idea of synthetics and substitutes in the food supply as replacements for traditional foods and food ingredients. However, unlike other consumer goods, synthetics and replacers in the food supply may have profound and widespread effects on biological processes involving human health. In recent years, with enlarged understanding of what contributes to optimal nutrition, we have a better appreciation of the complexity of the factors that contribute to good health. All cells of the body must be supplied daily with all necessary nutrients, in balanced states, and must be able to utilize and assimilate them. We lack information about the novelty foods and replacers. Rarely, if ever, are they nutritionally equivalent to the traditional foods or food components

5

that they replace. Nevertheless, consumers are now, unwittingly, taking part in what must be viewed as a novel experiment, the results of which may be fully comprehended only after a lag of ten, twenty, or even thirty years.

There is growing recognition that human health is related to one's lifestyle, including food choices. At a symposium on nutrition and cancer held in 1976, Dr. Ernest Wynder, president of the American Health Foundation, reported that most human malignancies are related to "our civilized way of living." Wynder and other participants discussed studies showing that cancer is more closely related to dietary factors than to other commonly publicized causes, such as smoking or carcinogenic industrial exposures. One participant, Dr. Kenneth Carroll, a biochemist at the University of Western Ontario, outlined a study showing that animals fed synthetic foods "tend to develop increased tumors," whereas control animals fed on natural foods developed significantly fewer tumors. The implications of such findings make it urgent to examine the impact of the new foods on human health.

Another aspect of dietary factors, which deserves attention but has been ignored, is the narrowing base of our foods, resulting from the extensive use of imitation and synthetic ingredients. Nutritionists have repeatedly emphasized the value of choosing to eat a wide variety of foods in order to obtain all necessary nutrients. But the extensive use of imitation and synthetic ingredients narrows the base. For example, if an individual chooses to eat fake chicken from textured vegetable protein on Monday, fake tuna fish from textured vegetable protein on Tuesday, fake ham from textured vegetable protein on Wednesday, fake turkey from textured vegetable protein on Thursday, ad infinitum, the nutrients will be far more limited than they would if they were derived from real chicken, tuna fish, ham, and turkey. Similarly, if an individual chooses to use nondairy creamer not only in coffee or tea but also with fruits and cereals and in cooking, the nutrients will be more limited.

Similar basic constituents appear in many guises in the new foods. Highly saturated coconut oil is not only in nondairy whipped toppings and creamers but also in imitation cheeses, imitation nuts, and imitation chocolate. Modified food starch is a constituent in many imitation foods. Any individual eating many of the imitation foods is eating an undesirably high level of saturated fats and an inferior form of food starch (and too much salt).

Fast food service further narrows the base. Such establishments serve food not only on university and college campuses but also in many high school and elementary school cafeterias. The limited choices of hamburger, French fries, and shake (sometimes without any milk base) are enjoyed by many children day after day, not only for the school lunch but later in the day, at home, or at a fast food service restaurant. When the fast food service took over the school cafeteria in one school, a mother groaned, "How can I send Johnny out for the Big Mac for supper when he'll already have eaten it at lunchtime?"

The New Look in Food

A study by the United States Department of Agriculture, released in 1972, showed that synthetic or substitute foods made up about 21 percent of all retail citrus beverage purchases, margarine had more than two-thirds of the table spread market, nondairy coffee creamers had about 35 percent of the light cream market, and substitute toppings had captured more than half the whipped topping market. Future substitutions of more vegetable protein for animal products were predicted, as were far-reaching changes in food selections. The first significant use of textured vegetable protein was as an extender with red meat and poultry in institutional cooking. The projection was that by 1980 about 20 percent of the red meat would be replaced in processed foods by vegetable proteins. Dairy products, too, were expected to continue being replaced by substitutes and synthetics.

The Vocabulary of the New Look in Food

Simulated foods are designed as total replacements for other foods. They are made to look, taste, and feel like the foods they replace. Examples include meat analogs, made from plant proteins to replace meat, or soy milk, designed to replace cow's milk. Meat and milk are whole foods, whereas meat analogs and soy milk are fragmented sets of extracted molecules representing only fractions of the original foods.

Synthetic foods are those made in a laboratory from nonfood sources, not from the products of field, pasture, orchard, stream, lake, or sea. Although great attention is given to approximating appearance, taste, texture, and other characteristics of real foods, the subject of nutritional value is ignored. Examples of synthetic foods include fruit drinks and nondairy creamers.

Imitation foods are those made to look like and replace foods for which standards have been established, but the imitations do not meet the standard requirements. An example is imitation mayonnaise, which contains less oil than is found in standardized mayonnaise.

Some products are "fabricated," "engineered," or "formulated," which means that they have been made from basic foods that have been restructured by taking them apart and then putting them back together in a new form. Their molecular structure has been rearranged. Examples include breakfast pop tarts and extruded (spun vegetable protein) snacks. The important question of nutritive losses that occur with molecular rearrangements of food components is one that is little explored or totally ignored.

"Artificial" designates food flavors and food colors that are made from both synthetic and natural ingredients.

Convenience foods require less labor in storing, handling, preparing, serving, or eating than foods they replace, since preparation of such foods was relocated from the kitchen to the factory.

Sometimes distinctions are made between traditional convenience foods such as bread and the more highly contrived new convenience foods such as frozen pizza or other frozen entrees.

The new convenience foods, created by food technologists, have become infinitely more sophisticated in recent years. For example, a continuous loaf of bread can be baked in a continuously heated oven, thus providing an everlasting supply of wrapped slices, each identical to the next one.

In England a commercial veal-and-ham pie is available with a hard-cooked egg in the middle of it. Some slices inevitably would contain more white than yolk, owing to the shape of the egg. But veal-and-ham pie manufacturers can be supplied by a continuous "rod" of hard-cooked egg made in another factory as a tube of white with a core of yolk right through it.

In anticipation of the future, Dr. Magnus Pyke has wryly suggested an additional term. "It seems to me probable," he wrote, "that together with the convenience foods of advanced technical food manufacture . . . there will appear on the market 'inconvenience' foods. These may well be saddles of mutton or perhaps half sheep for the family deep-freeze, to be brought out on birthdays or other community festivals."

2. Restructured Animal Protein

《《《《《《《《

The complex structure of protein in its unaltered form . . . is essential for biological life. The immense complexity of the protein molecules . . . not only affects the texture and characteristics of the living tissues from which they are derived but is also an integral part of the process of life. . . . A comparatively moderate degree of heat—for example—only enough to warm milk a little above blood heat—may so disrupt the chemical configuration of a complex protein molecule—for example, the enzyme, rennet—that it will no longer carry out its biological function—that is, to coagulate the milk protein.

—Dr. Magnus Pyke, *Food Science and Technology*
(London: John Murray, 1964)

Extruding technology is the answer to food production in the future.

—*Canner/Packer,* January 1976

Ask the seasoned extruder man . . . [our extruder machine] can be run one day at 6,000 pounds per hour as textured vegetable proteins and the next day at 20,000 pounds per hour on typical medium density/cereal/protein mixtures. Producing finished products in almost any shape you can imagine . . .

—advertisement, *Food Processing,* April 1976

Restructured Animal Protein

The polymer, nylon, is made artificially by purely chemical means in a chemical factory. Cold-drawn in a particular way, this *physical* operation causes a rearrangement to occur in the *chemical* configuration of the nylon molecules. In just the same way, the long complex molecules of protein in foodstuffs may be changed by physical treatment.

—Dr. Magnus Pyke, *Food Science and Technology*

Controlled particle sizing through compaction: some [food] products are not available or produceable in the particle size, bulk density, or flow characteristics required in their final use or further processing. Several methods of agglomeration are available. . . . Sub-size product is force fed between two grooved, pressured rolls, discharging as a dense, corrugated sheet, which is then fed to a [granulating machine] equipped to yield the desired particle size.

—advertisement, *Food Processing,* December 1975

No longer is food grown or raised. It is developed in much the same way that Bethlehem Steel develops a metal formula for the particular needs of a customer.

—Nicholas von Hoffman, "Ersatz Food," Washington *Post,* 16 April 1971

A recent announcement that Sainsburys (London) are to test-market, through 30 of their stores, pies with textured vegetable protein fillings made by a textile company rated little more than a small paragraph in the London trade press. Two of the most significant aspects of this fact are probably the involvement in food production of a non-traditional supplier (in the textile or chemical industry) and the generally unsurprised acceptance of the development by the distributive trade and the consumer.

—Elrick and Lavidge, Inc., *Marketing Today,* 1975

If a particular fabricated food product is attractive, good-tasting, economical, safe and has good texture, mouthfeel and good shelf life is this enough? Nutritional scientists, generally, do not believe these factors alone are always sufficient in designing a new food today—especially when the food resembles or imitates a traditional nutritious food and is going to be used in significant amounts.

—Dr. George M. Briggs, Department of Nutritional Sciences, University of California, "Nutritional Aspects of Fabricated Foods," in *Fabricated Foods* (Westport, Connecticut: Avi, 1975)

〉〉〉〉〉〉〉〉〉

Formerly, the restructuring of foods involved using waste foods or foods of inferior quality and making them appear more attractive or palatable by changing them into new forms. "Formed fish," consisting of molded scraps of dried fish, was consumed throughout the Middle East in ancient times on long sailing voyages. During the Middle Ages, serfs and peasants gathered foods left from the sumptuous feasts of noblemen and boiled and softened such scraps, then forced them through a small orifice of a funnel-shaped vessel. The resulting paste, shaped into patties, balls, or loaves, was recooked. Of course, sausages have long been recognized as restructured meat. However, many of the newly created restructured foods are not always recognized as such by consumers.

Modern restructuring of food has been inspired by the techniques and equipment of nonfood industries. The spinning of vegetable protein is similar to the spinning of nylon. Essentially, the process of extruding food from an extrusion machine does not differ from extruding industrial products such as metal, plastic, or rubber.

With ever-mounting pressures for mechanization, uniformity, quality control, and maximum efficiency—goals that now pervade the highly industrialized sections of the world—food production has inevitably felt the impact. Although such goals may have value for certain consumer goods, they may be highly inappropriate for food production. Despite the great technological skills that have been developed in restructuring foods, basic questions have been ignored. What are the nutritional values of restructured foods compared to their traditional counterparts? What happens to nutritional values when molecular structures of foods are rearranged, destroyed, subjected to extreme heat or pressure? How severely is protein denatured? How drastic are the nutrient losses in foods that have been cooked, frozen, thawed, and then reheated several times?

Restructured Fish and Sea Food

Early production of frozen fish sticks demonstrates how nonfood technology was applied to food processing. Wrapped blocks of frozen fish were tempered, sawed into pieces lengthwise, and fed into a chopper that cut the individual fish sticks to a predetermined size, as lumber might be cut in a woodworking shop. The uniform fish sticks were dropped onto a conveyor, battered, breaded, and fried. By means of a transfer belt they went into a tunnel, were refrozen, and then were packaged, heat sealed, stored, and shipped. Ultimately, they were thawed out and reheated.

Present methods of cutting the fish have refined the art. Frozen blocks of fish are sawed diagonally, to suggest the shape and look of natural fish fillet. Pieces that fail to fit into the special forms are saved and re-formed into a mold that is later recut into other portions.

Extrusion machinery, adjusted to produce a crescent shape, can restructure some three to four hundred tiny thawed shrimp into sixteen to twenty large "shrimp" or shape them into rounds or "cutlets." Clam strips are also extruded, into strips.

Recognizing that the public is not always aware that such products are restructured, the Food and Drug Administration proposed regulations for labeling products resembling fish sticks, fried clams, and breaded composite shrimp products. Qualifying phrases that are distinctly different from other printed or graphic matter and appear "in legible boldface type" must be at least half the size of the name of the product. Such proposals may put retail shoppers on guard, but this information does not appear on restaurant menus.

The Eight-Inch Restructured Egg

A Danish food engineering company devised machinery to restructure eggs into a salami shape about eight inches long. The machine, operated by one person, can form up to five hundred restructured eggs hourly. Such reconstituted eggs have been available commercially in Europe since 1964 and currently are used by restaurateurs in the United States. They have recently been introduced in retail stores.

Why use restructured eggs? They have the built-in features of portion control and uniformity. Every slice of the reconstituted egg contains the same proportion of white and yolk, and the problem of yolkless end slices is eliminated. Reconstituted eggs are used in slices in salads or as garnishes.

Restructured eggs are frozen. Even after having been stored frozen for two years, they are reported, after being defrosted, to have the same taste and texture as eggs freshly cooked.

Eggs are also reshaped into squares. In England, hard-cooked eggs are shelled and placed in small plastic boxes. Gradually, the tops of the boxes are gently pressed down so that the eggs are slowly pushed into the corners of the box. Why square eggs? Presumably, sliced square eggs are convenient for sandwiches, and they slip off platters less easily than ovoid ones.

Restructured Meats and Poultry

Ready-to-eat foods such as pressed beef, chicken, or turkey "loaf" or "roll" are familiar convenience foods popular with many consumers and widely available in retail markets. One may choose, deliberately and voluntarily, to use processed meats or poultry in lieu of unprocessed ones, but now such fabricated products are being palmed off as the real thing. Increasingly, those restructured meats and poultry are served in restaurants and other feeding institutions and listed on the menu as sliced beef, chicken, or turkey, in sandwiches and other dishes. The use of such products is favored because of their low cost, uniform composition, portion control, savings in labor, ease of handling, and long shelf life. Restructured meats and poultry consist of low-quality proteins, and the products contain objectionably high levels of fat and salt, chemical additives, and water. Their flavor is usually poor.

Restaurants and other such institutions now serve a variety of restructured meats that are somewhat unfamiliar to the public. "Roast beef" may be fabricated by combining chunks of beef with salt, fat, and vegetable protein and stuffing it into sausage casing. It is then heated and transformed into a solid piece of "meat" with fat rippled through it. Beef may be restructured so that it closely resembles the color, appearance, bite, and flavor of whole-muscle steaks. Such products, inaccurately listed as "steak" on menus, are especially favored by fast food chains. The *New York Times* food writer Mimi Sheraton observed that the practice "permits the use of cheap, tough cuts, otherwise inedible when broiled." (Nor are steaks, real or restructured, necessarily broiled anymore. Restaurants may use special equipment to brand grill marks onto the meat and make it appear broiled.) One "beef grill steak" product sold to restaurants consists of shredded, chopped portioned beef delivered frozen. After thawing, and with the addition of gravy or other sauces, the product is served as Swiss steak.

Restructured "steaks" were recently introduced into retail markets. When freshly ground lean beef was selling for about $.99 a pound, restructured steaks were selling for $1.50 to $1.60 a pound. Consumers have noted that restructured steaks tear apart, rather than cut like a whole-muscle steak. Restructured beef strips and pork breakfast strips have been introduced at retail stores. Such products consist of cured chopped and formed pieces of meat with smoke flavor added, and several chemical additives.

Veal is also being restructured into veal "steaks" and lamb into lamb "chops." Re-created and restructured pork produces ham "steak" that is charbroiled, grilled, barbecued, baked, or used for Hawaiian-style ham. Marbling effects, if desired, can be created with fat. Restructured pork products also include "choplets" and Canadian bacon–like strips.

Extended Restructured Meat Products: Less Meat and Poorer Quality

It is now possible to use less meat and of a lower grade in products such as processed beef, corned beef, pastrami, and ham rolls by adding rehydrated textured vegetable protein. That extender may replace more than 20 percent of the meat. Extended hot dogs are in production; extended bologna and Polish-style sausages will be used in the federally supported school lunch program. Extended salami, smoky links, and Italian-style sausages with additional extended meat, poultry, and fish products are being developed.

Mechanically Deboned Meat: Turning Garbage into Money

In traditional manual meat deboning even a skillful butcher cannot retrieve all the meat from the carcass. Mechanical deboning equipment can salvage this meat, but it is of low quality and bone solids may be present. Formerly, the USDA permitted mechanically deboned meat (MDM) solely for use in pet food or for export and restricted bone solids to up to 1 percent.

Meat processors, claiming that nearly a billion pounds of meat could be retrieved annually by mechanical deboning, pressed to have restrictions relaxed. The argument appeared persuasive, in times of high meat costs and concerns about global needs for protein.

In April 1975, without advance public review, the USDA relaxed its restrictions. The agency issued a temporary regulation, given to federal meat inspectors, that would allow processors to use MDM in an extensive array of food products intended for human consumption, pending public consideration of a permanent regulation. Consumer groups protested and declared that the procedure not only was illegal but also showed what a sham the USDA's consumer representation plans are.

No provision was made to label MDM specifically, so, at the point of purchase, consumers had no way of knowing if the product contained MDM. The USDA admitted that a small amount of finely ground bone would remain in MDM but claimed that consumers would not be able to detect it, and besides, there was a nutritional bonus in calcium from the bone. Rodney E. Leonard, a former USDA official, charged that the ground bone in meat products is hazardous and should not be passed off as something beneficial to consumers. Due to the finely ground nature of MDM, it is highly susceptible to rapid quality deterioration. The USDA's proposed regulation failed to include any standards for bacterial content. Other problems involved sharp bone fragments and contamination of bones with lead, strontium 90, and fluorine. Also, some individuals suffer from hypercalcemia and must avoid calcium.

The agency had proposed a maximum fat level of 50 percent for processed meats made from MDM, which is far higher than the already undesirably high level of 30 percent maximum fat in ground beef. (Frankfurters are specifically covered under separate regulations.) Consumers objected to that high level of fat, especially in view of data that suggest the undesirability of excessive fat consumption. The use of MDM was termed by one consumer group a procedure for "turning garbage into money." The large meat packer Oscar Mayer decided against the use of MDM in its products. Its decision was based on its own studies that convinced the company that MDM could not be used "without changing the characteristics and perhaps lowering the quality of its products." The company recommended that MDM be appropriately labeled as "mechanically deboned beef," "mechanically deboned pork," and the like, which the USDA's regulations had failed to require. Oscar Mayer further recommended that MDM be considered a meat by-product rather than meat.

Because of various criticisms of MDM, the USDA suspended MDM manufacture in the autumn of 1976, pending further investigation. The USDA's Food Safety and Quality Services then conducted an extensive review of MDM, with input from university scientists, an interagency panel, public health officials, and consumers. The resulting new set of proposed regulations, issued in October 1977, would abandon the term *mechanically deboned meat* and use instead "tissue from ground bone" (TFGB). This wording would have to be included in the list of ingredients contained in the meat product so that persons who must stringently restrict calcium intake could avoid such products. TFGB could be used only up to a maximum of 20 percent of the meat used in a product and could not be used in strained baby, junior, or toddlers' foods, because of its fluoride content.

Restructuring TFGB

TFGB is pulpy in appearance and describable as an emulsion. To give it some structure, which is necessary to convert it into a salable food product, a special structured protein fiber (SPF) is added. SPF, composed of a series of microscopic hollow tubes, has an ability to absorb and retain water and fat by capillary action. Added to the salvaged meat, SPF bestows a meatlike structure and gives both the "mouthfeel" and the "biteability" of red meat. SPF can also be combined with mechanically deboned poultry to be shaped as "legs" or "drumsticks." Combined with mechanically deboned fish, SPF permits the product to be restructured as small whole "fish" or "fillets."

Cooked Beef Fat Tissue Solids: Extending Meat Products Further

Cooked beef fat tissue solids (CBTS) is an end product made from edible beef fats that have been rendered with high heat. The product consists of particles with a meaty texture that are beige colored and bland in flavor. CBTS is blended with many restructured meats. Although rendered CBTS no longer contains much fat, it has an ability to absorb and retain both fat and water. It also adds structure to foods and acts as a binder, stabilizer, emulsifier, and thickening agent.

Since CBTS costs less than one-fifth the price of beef, its low cost makes it especially attractive to processors as a meat extender. It can replace from 20 to 26 percent of ground beef without being detectable by taste. Although USDA regulations require the name "cooked beef fat tissue solids" to appear in the ingredient statement of processed foods to which it has been added, the term is apt to be unfamiliar to most consumers. The USDA permits this extender in hot dogs and cooked sausages at levels up to 15 percent. CBTS is found in prepared cooked frozen meat products such as stews, soups, potted meats, loaves, chilis, tacos, sloppy joes, and spaghetti sauces.

Quality Protein or Technological "Progress"?

The case of cooked beef fat tissue solids, fought by the state of Michigan, is an example of the many legal battles between state and federal agencies about food quality standards. Frequently, states have chosen to establish standards that are higher than federal ones and more beneficial to consumers. This is especially true for meat and poultry regulations.

Since food products may be produced locally but be transported to other states, they may be regulated by both state and federal requirements. When the regulations differ, state standards repeatedly have had to be lowered to accommodate lower federal standards.

The Michigan Department of Agriculture (MDA) requires that food products containing ingredients that are federally approved but disapproved by the MDA be specifically labeled and prominently identified in retail stores. In effect, this law permits the sale of products containing such ingredients but puts consumers on guard as to the product's inferiority, according to state standards.

The MDA had a long controversy with the USDA over conflicting rules about the use of animal by-products in processed meats. The MDA contended that CBTS was a "slaughterhouse by-product" that should not be used in Grade #1 sausages. A Michigan state court ruled against MDA, declaring CBTS "a wholesome and nutritious product" at least equal in quality to the "raw materials that would otherwise be used in Grade #1 sausage." This ruling, and others that lower the standards, do not serve the best interests of consumers. Such decisions equate low with high quality and officially encourage food debasement.

3. Animal Proteins versus Textured Vegetable Proteins

《《《《《《《《

Most of the "rave notices" for the new EPP [Engineered Protein Products] come from the manufacturers themselves and from some more adventurous food writers.

—*The Many Faces of Engineered Protein Products* (Chicago: National Live Stock and Meat Board, August 1974)

The period is not too far in the future when many people will prefer soybean products to "meat," says . . . [a] senior research chemist for . . . one of the world's largest soybean processors.

—Washington *Post,* 4 December 1975

[A textured vegetable protein product] will let you save as much as 50 to 60 percent on protein costs in many finished products—coffee whiteners, whipped toppings, snack dips, imitation cheese spreads, protein-fortified beverage powders, baby foods, and others.

—advertisement, *Food Processing,* May 1976

Ham it up for 1¢ an egg with *new* General Mills Protein 11 ® Crumbles. They look like ham. Eat like ham. Taste like ham. But at 1¢ a serving, they sure

Animal Proteins versus Textured Vegetable Proteins

don't cost like ham. Now you can add ham-like flavor and texture to scrambled eggs, omelets, pancakes, waffles, biscuits, casseroles, western sandwiches—for a penny.

—advertisement, *Institutions/Volume Feeding,* July 1975

Foods made from textured soy protein (spun fiber) cost more than items containing natural meat products.

—A. C. Peng, "Plant Proteins in Foods," *Ohio Report* (Wooster, Ohio: Ohio Agricultural Research and Development), September–October 1973

A meat analog: an engineered protein food product meant to look and taste like meat. . . . Manufacturers claim a meat analog is not necessarily an exact duplicate for a specific cut of meat but a "new protein food of controllable quality which is compatible with many traditionally accepted meat dishes." To most thoughtful persons the statement supports itself. It is, nonetheless, also a euphemism intended to avoid inflaming or irritating traditionalists or competitors.

—*The Many Faces of Engineered Protein Products*

[After sampling soybean-based meat substitutes] one participant . . . conceded the products were probably the wave of the future. . . . I guess people will come to like them, he sneered. . . . My dog already has made his peace with Gaines-burgers. . . . But every once in a while, when he sniffs the air redolent of rare beefsteak, I can see passing through his mindless eye the distant memories of what used to be.

—*Forbes,* 1 April, 1973

I recently yielded to an impulse to try the textured vegetable protein product referred to as "breakfast strips." . . . As an experience in eating, my own conclusion is that these strips are to bacon as Masonite is to solid walnut, or as marbelized linoleum is to marble.

—Craig Claiborne, "Certainty, and Doubt About Names,"
The New York Times, 25 April 1976

Vegetable protein foods . . . provide a protein source equal to any available today—including meat.

—advertisement, *Health Food Retailing,* March 1972

Because of essential amino acid imbalance, most plant proteins are less efficient, have lower digestibility, and a lesser biological value than animal proteins. A common measure of a protein's value is its protein efficiency ratio (PER) or its net protein utilization (NPU). For example, soy flour is higher in crude protein than meat, but has a lower PER and NPU. Another shortcoming is that the level of all amino acids may be too low to meet body needs.

—"Synthetics and Substitutes for Agricultural Products, Projections for 1980," *USDA Marketing Research Report No. 947* (Washington, D.C.: USDA, March 1972)

〉〉〉〉〉〉〉〉〉

A wide range of sources has been suggested to extend or replace traditional sources of protein, such as meat. The list of alternatives includes protein from plants, such as cottonseed, peanut, alfalfa, tobacco, and leaf; and animal by-products, such as fish meal, blood, nails, skin, hair, hooves, feathers, and wool. Further, suggestions have been made to use insects, fungi, bacteria, yeast, algae, single-cell protein grown on a petroleum base, and even sludge. Some of these sources present obstacles, such as the presence of toxins (gossypol in cottonseed), the danger of heavy metal contamination (in sludge), high energy demands (single-cell protein), and human attitudes toward food (insects). Most of those protein sources are not yet developed as human food sources. A few (cottonseed, peanut, and alfalfa) sources may become technically feasible, but most of them are merely being discussed. None offer nutrition equivalent to animal protein from traditional sources.

To date, the soybean is the sole challenger of animal protein. It is being used increasingly not only as an extender of meat but as a total replacer.

The idea of meat substitutes is not new. Dr. John Harvey Kellogg is credited with being the first to use plant protein to create an early vegetarian meat substitute, peanut butter. Kellogg and his brother, W. K. Kellogg, created the first dry breakfast cereal. Dr. Kellogg then developed the first meat substitute by mixing soy meal with nut butter and cereal.

George Washington Carver studied the protein content of legumes, such as the peanut and soybean, in order to exploit their usefulness. Ultimately, the contributions of Kellogg and Carver led to the creation, about 1910, of a pioneering vegetarian food company that made "meat without bones" products. A meatless loaf was created with peanuts as the base and with wheat and soybean flours added. A succession of meat analogs from grains and legumes followed, with products shaped to simulate meat products such as veal cutlets, burgers, wieners, and breakfast sausages. Those items were marketed principally to vegetarians.

Although soybean flour and grits have been commonly used to enrich breads and other baked goods, soybeans were unpopular in the form of meat analogs during World War II. Soybean products were described as "beany, bitter, and loaded with indigestible carbohydrates." Soybean analogs were sufficiently unappetizing for the word *soy* to become a disparaging epithet. If things went according to plans in a wartime battle, the code words were *pork sausage;* if disaster impended, the term used was *soya link.*

The breakthrough for using soybeans as meat substitutes came in 1957 when a textilelike process was perfected for spinning edible soy fibers that could be shaped into meatlike products. Flavor chemists and seasoning manufacturers combined technological skills to create stable flavors that imitated those of various meats. Soy protein isolate, a highly processed fraction of soybean used in many processed foods, has been available in the United States since 1960. By 1970 its annual production reached some twenty thousand metric tons.

The 1970s became a propitious time for public acceptance of meat extenders and replacers. American consumers, as well as restaurants, hospitals, nursing homes, schools, and other institutions, faced rising food costs and were searching for less expensive food alternatives. Also, to many individuals soybean analogs appeared to be the solution to the saturated-fat-and-high-cholesterol issue. There was also growing concern about world population, inequities of distribution and consumption, and predictions of future food shortages on a global scale. The lowly soybean, long underutilized, seemed to be the panacea. Although alternate protein foods are now being investigated, the soybean is far in advance of them all in technological development and market dominance.

By the mid-1970s soybean flour and grits, concentrates, isolates, and extended or spun textured products were in such widespread use that an FDA dietician and nutritionist observed: "Today, chances are good that one or more food items in the average family's weekly grocery supply contains some form of plant protein product. And chances are even better that tomorrow's groceries will have many more foods made partially or completely from plant protein." Spun protein, which can be fashioned to look like whatever kind of product it is intended to resemble, now simulates beef, ham, luncheon meats, pepperoni, turkey, or chicken slices; bacon or bacon bits; pork sausage links or patties; Salisbury steak, or fried chicken. It is added as an extender in pressed beef, corned beef, pastrami, or ham rolls, is shaped to resemble fish fillets, scallops, or veal cutlets, appears in chunk form imitating ham, chicken, or tuna for salads, and is added to flaked seafood. In addition, the textures of such foods as shredded coconut, dried fruits, and nuts can also be imitated. To compound the illusion, a processor of one chickenlike product made from soybean protein whimsically included a plastic wishbone in the package.

Textured Vegetable Proteins as Meat Extenders

In the early 1970s, the introduction of blends of ground meat and textured vegetable protein in food stores was launched by their processors with a vast promotional campaign. By the summer of 1973 one out of every four food stores across America sold such meat–soy protein blends. Sales have often outstripped the sale of pure ground beef. Combined with meat protein and used in limited quantity, soybeans may be an acceptable extender, but there are drawbacks.

Among those who decided not to sell the blends was one major food distributor. A meat supervisor, speaking for the company, explained the decision: "Certainly it's cheaper than pure beef, because water is cheap. The soybean can absorb up to four times its weight in water and retain it. In the end, these products are 20 percent water. We don't believe in additives of any kind. We sell pure beef. Otherwise we open an area of doubt for the consumer. The housewife will lose confidence in what she buys."

State consumer protection agencies consider excessive extenders or water in meat–soy protein blends to be a form of economic cheating. By law such extenders are limited to 25 percent of the blend, and any with an excess of extenders or water is considered adulterated.

Excessive fat may be another problem encountered in beef–soy premixed products. By law, ground beef may not contain more than 30 percent fat, which many consumers consider too high. The fat limit in beef–soy premixed products is not regulated, nor does the consumer have any choice or control over the quality of the meat used in such premixed products.

Although the addition of a commercial soy extender, when blended with meat, *may* lower the total cost for protein, it does not necessarily do so. Some textured vegetable protein products sold in expensively packaged four-ounce foil pouches cost *more* than the ground beef they replaced.

Diabetics, and others who must restrict their carbohydrate intake, need to know that textured vegetable protein products contain carbohydrate, which is not found in the animal products they replace.

Introduction of the meat–soy blends left consumers confused in selecting food both in stores and from restaurant menus. At the retail level, the names for these meat–soy blends varied from place to place. Terms such as *burger, superburger,* or *juicy burger two* were prohibited in some states. *Hamburger, ground beef,* and similar designations were reserved for the traditional all-beef product.

The Consumer Protection Division of the Los Angeles County Department of Health Services launched a "Truth in Menu" program for food operations and was especially concerned about hamburger versus imitation hamburger. California law requires that if imitation hamburger is sold or served in a restaurant of any type, a list of ingredients must appear on the menu, unless it contains no more than 10 percent added protein and water and contains no binders or extenders. Operators may not refer to their products in such words as *hamburger* or *burger*, but must refer to them as *imitation hamburger* or by any other term that accurately informs the customer of the nature of the food product being served or sold. Other states concerned with the same problem have also taken action.

In places where control is lacking or lax, there is a strong temptation to use soy extenders without saying so on the menu. Upon analysis, "pure beef" hamburgers from some fast-food outlets have been found to contain not only soy as an extender but starch as well and to be above the fat limit, too.

In 1973 a trade journal poll revealed that soy extenders were being used by 6 percent of the restaurants surveyed, more than 20 percent of the hospitals and nursing homes, nearly 30 percent of the colleges and universities, and more than 50 percent of the schools, after textured vegetable protein products were approved in 1971 for the school lunch program.

Recipes were devised (especially for institutional cooking) to use soy extenders in dishes such as codfish cakes, sardine salads, chicken croquettes, cornmeal mush, cornbread, and oatmeal cookies, and even in such unlikely dishes as Waldorf salad, flavored rice, and scrambled eggs.

Later, textured soy protein products were introduced at five to six cents per serving as total replacers for meat in dishes such as tacos. The meatless mix is composed of textured soy protein, flavorings, and colorings. Other formulated meatless mixes include chili and sloppy joe. This use of soy protein is referred to as second generation.

Meat Analogs: Simulated Pork Links, Pork Patties, Bacon, and Others

Extravagant, inaccurate claims are made for meat analogs. Consumers are told that these products are less costly than the meat they replace, are of equal or even superior quality, taste as good, and are more healthful.

Animal Proteins versus Textured Vegetable Proteins

It is costly to texture and shape meat analogs into forms accept- able to consumers. As a result, some of those products are more ex- pensive than the animal proteins they replace, such as red meats, fowl, or fish.

One simulated chicken product, spun from field beans, has been promoted in England, where broilers are more costly than they are in the United States. The product was more expensive on a per- pound basis than chicken or trout and contained more than 60 per- cent of its calories as animal fats, which is about three times as much fat per ounce as pure poultry meat.

"High quality protein," "complete protein," "a protein source equal to any available today—including meat," "vegetable protein foods can adequately replace meat," "richness of protein," "full bio- logical value"—these are commonly used but untrue claims for meat analogs. One food writer described these products as "today's wonder foods—they contain twice the protein of meat or fish, three times the protein of eggs and eleven times the protein of milk." Such statements mislead. *Meat analogs are inferior in nutrient qual- ity to protein from traditional animal sources such as meat, poultry, fish, eggs, milk, and cheese.* Although the soybean has a higher nu- tritional value than other legumes, it has less nutritional value than animal protein. Soy as a meat *extender* may in limited amounts com- bined with animal protein be acceptable, but soy cannot be consid- ered nutritionally equivalent or superior as a meat *substitute.*

The drastic alkali treatment of processing soy into soy products reduces the net protein value. In producing soy protein isolates, for example, three amino acids (lysine, serine, and cystine) are reduced, with cystine the most sensitive of the three. In addition, the severe alkali treatment produces an amino acid derivative, lysinoalanine, present in the treated protein. Lysinoalanine is toxic. This substance has been shown in experiments to be detrimental to the rat kidney. Hence, the safety of alkali-treated soy products needs investigation.

The USDA reported that "because of essential amino acid imbal- ance, most plant proteins are less efficient, have lower digestibility, and a lesser biological value than animal proteins." Soybeans are low in an essential amino acid, methionine.

Attempts to add methionine to a meat analog are unsatisfactory since such an addition is potentially toxic. Methionine has been known to combine with other constituents and form toxic compounds. Animal protein foods contain *all* the essential amino acids and are present in an ideal balance for human consumption.

The total nutritional offerings of the meat analogs depends in part on their content of added fat. In experiments with animals, any changes of fat and nonnutrients in food intake can alter the activity of enzymes that metabolize drugs and carcinogens.

Rehydrated soy protein concentrate contains less available phosphorus, riboflavin, niacin, and Vitamins B6 and B12 than meat.

Soy protein isolates are loaded with sodium, whereas fresh meat contains very low levels of sodium. The high sodium level in meat analogs is undesirable for all persons, but especially for those who must restrict their sodium intake. Madeleine Saari, a nutritionist with the Montgomery County (Maryland) Heart Association, said that she would not allow persons on even a mildly restricted sodium diet to eat products such as meat analogs that simulate pork patties. Such products contain from 600 to 800 or more milligrams of sodium in a single serving.

Although calcium and zinc are found in vegetable proteins, these minerals are rendered *less* available, because of the presence of oxalic or phytic acids, which form insoluble salts. Meat, which contains neither oxalic nor phytic acids, is a reliable source of these minerals. The increased substitution of vegetable protein for meat in the diet is thought to contribute to zinc deficiency, which is widespread in some segments of the population.

The low availability of zinc in vegetable proteins, contrasted with the high availability of zinc in animal proteins, was demonstrated in experiments with chicks and rats (see chart, page 29).

Although soybeans contain iron, it is also bound up in a form of low availability to the body. The iron in meat is in a form that is well utilized.

Soy protein is poorer in trace elements such as chromium or selenium than are animal protein sources.

It has been suggested that missing nutrients in meat analogs could be supplied by fortification. Critics point out that this approach is impractical because fortification with certain nutrients may be too

Animal Proteins versus Textured Vegetable Proteins

PERCENTAGES OF ZINC AVAILABILITY

	WITH CHICK ASSAYS	WITH RAT ASSAYS
High-lysine corn	65	55
Control corn	63	57
Rice	62	39
Wheat	59	38
High-lysine corn germ	56	——
Control germ	54	——
Sesame meal	59	——
Soybean meal	67	——
Egg yolk	79	76
Fish meal	75	84
Oysters	95	——
Nonfat milk	82	79

expensive and not enough is known about balancing various nutrients. Some, such as the amino acids, can be toxic, if oversupplied.

Not all nutrients are yet known. At present more than fifty essential nutrients have been identified, and it is suspected that the total number may run as high as seventy-five to one hundred if identification is ever completed. In addition, there are still many undiscovered interrelationships among those nutrients.

It has been observed that "technologists are producing meat substitutes faster than regulating standards can be devised for them. A new product often precedes proper regulation for it." A similar statement could be made about the lack of research on those products. Meat analogs have not been subjected to long-term feeding tests. The health implications of the relatively low availability of some of the soy nutrients in the digestive process have not been thoroughly explored.

In a critical appraisal of new meat analogs, made from legumes as well as other sources, an editorial writer in the British medical journal *The Lancet* raised several questions. Do we need such products? In affluent countries, populations are not protein deficient, and in developing countries people cannot afford the cost of such products

29

as meat analogs. Protein-deficient populations need to grow and eat legumes directly rather than to import expensive, highly processed forms of them. Such people should be taught how to combine various vegetable proteins to maximize protein quality. Thus, it appears that meat alternatives are being produced "for no sound nutritional reason."

Furthermore, *The Lancet* warned that the introduction of any new food "must be regarded with qualms, scepticism, and even suspicion." Is it safe? What will it do to the population over a period of time? Allergies have already cropped up in people eating some novel protein foods. The problems are similar to those of introducing a new drug. At least for new drugs there is a trial period during which drugs are administered under special medical supervision. Will it become necessary to sell these new foods by prescription? Will epidemiological work be done? Such foods, widely introduced, may contribute to causing diseases that appear only after long latency periods. *The Lancet* urged that a committee be established to monitor the effects of new foods on patterns of illness in the population.

The Politics of Textured Vegetable Protein

The critical issue of whether the new vegetable protein products are nutritionally equivalent to traditional sources of animal protein was hotly debated in 1971. The USDA gave its official blessing to alternate foods for child-feeding programs and allowed up to 30 percent textured vegetable protein as a meat replacer in dishes such as meatloaf, chili, and sloppy joe. This percentage was higher than the allowance in ground beef–soy protein blends sold at the retail level, which was limited to 25 percent.

Approval came in February 1971, at a time when schools were struggling with problems of meeting increasing food costs. With the use of textured vegetable protein in the school lunch program the ingredient cost for meat dishes could be reduced by as much as 20 percent.

The processors of textured vegetable protein products were elated by the USDA's regulations. The policy was viewed as an opening wedge for the introduction of many more alternate ingredients.

Critics charged that under the guise of good nutrition, food corporations were pushing alternate and junk foods into schools and other child-feeding programs. "Several USDA officials seem to have a lot to gain, while moving freely in and out of these companies," charged Susanne Vaupel, one such critic. "One of the manufacturers of textured vegetable product is the Ralston Purina Company," she said, noting that Earl L. Butz, who at the time was secretary of agriculture, had served earlier on the board of directors of Ralston Purina. Butz's predecessor as secretary of agriculture, Clifford Hardin, now serves on its board. Ms. Vaupel named other USDA officials who formerly served in food companies that now supply "junk foods" approved by the USDA for the federal feeding program.

The Society for Nutrition Education expressed disapproval of the USDA's decision, stating that "it would seem . . . to be a poor national policy to permit or encourage the use of . . . alternative foods on the basis of cost." The society noted that when nutrient intakes were manipulated in the proposed manner, there was the danger of creating an imbalance or deficiency in some of the micronutrients.

The Food Research and Action Center submitted a statement signed by over one hundred professionals in the fields of nutrition, medicine, and education, as well as by representatives of more than a dozen citizens' organizations, viewing the USDA's policy as

> another step in a long and accelerating trend toward a more and more highly processed diet, consisting of "mock foods" concocted in the laboratory. . . . Clearly textured vegetable protein does not represent a move toward the rational use of vegetable protein in a world heading towards a shortage of animal protein. It is, instead, a triumph for the food technologists and the food companies. So long as the technologists had only vitamins and minerals to play around with, people still had to eat some "real food" to get their day's proteins. With textured vegetable protein, science has at last conquered "real food." . . . We consider this a development to be resisted. In a country with an abundant food supply, the continued proliferation of such products has no conceivable purpose other than allowing for the continued growth of food companies. If the American people, know-

ing the real options and the real risks, choose to take the path toward an increasingly synthetic diet, it is, of course, their privilege. To take them along this path without their informed consent is a serious deception. To allow their children to be taken along this path without their parents' knowledge and consent—as these regulations propose to do—is a nutritional crime.

The Soybean Is Not a Panacea

Although fabricated products such as textured vegetable protein cannot be equated with animal protein, there still remains the basic commodity, the soybean. Is this legume an answer to world hunger and shortages of quality protein? Admittedly, the soybean is of better biological quality than other vegetable proteins, but its drawbacks are overlooked by many of its enthusiastic boosters.

It has been argued that soybeans have long been used to sustain people in the Orient. Although this is true, we have little information about what varieties of soybeans were used, the genetic work done in developing varieties, or how the soybeans were combined with other proteins. We lack data about the health of these people. What we do know is that Orientals consider fermented soybeans and soybean products more palatable and more readily digestible than unfermented ones. Soybean products in the Orient have mainly been used for flavoring purposes, with the exception of two protein foods, tempeh and tofu (bean curd).

The raw soybean contains several antinutritional elements. It contains trypsin inhibitors. It is antagonistic to Vitamins A and B_{12}. It is a hemagglutinin that agglutinates the red blood cells. These undesirable features are largely deactivated by high heat, but unfortunately this treatment denatures the soybean's protein. The soybean, as well as flour and other products made from it, contains a goiterogenic principle that even after heat treatment is not completely deactivated. Thus, the use of the soybean as a staple in the diet is questionable.

4. Biddy's Shell Eggs versus Egg Substitutes

〈〈〈〈〈〈〈〈〈

New products appear on the market which are touted as the panacea of all our ills, as old established foods such as milk, eggs and the rest are condemned. The new Egg Beaters® . . . was severely criticized at a [1974] meeting of food technologists. . . . The major outcome was this: as newer foods are fabricated and old ones modified, the scientist had better know full well what the nutritional effects will be so that advertising will be truthful. They don't. Full research indicates many of the "great scientific food discoveries" just don't have it.

—letter to editor, *American Dairy Review,* September 1974

Egg Beaters® are a synthetic, processed food. . . . While Egg Beaters® may be similar to eggs in some ways, they are hardly an exact match in nutrition. Accordingly, we consider it deceptive for Standard Brands to advertise that Egg Beaters® offer "the nutrition of fresh eggs." . . . Distrust advertisements involving cholesterol. They're selling products, not health.

—*Consumer Reports,* March 1974

Egg Beaters®, the artificial eggs laid by the chemists . . . and costing about 50 cents more than hen's eggs . . .

—Colman McCarthy, "The Chemical Cuisine," Washington *Post,*
28 May 1974

33

Egg Beaters® are home on the range—are they on yours? The success of Egg Beaters® . . . in supermarkets proves you have many customers who must (or want to) limit their dietary intake of cholesterol. . . . Profit from this trend that lets you put a higher menu price on omelets, pancakes, french toast, scrambled Egg Beaters®. In fact, you can just thaw and use Egg Beaters® for any dish requiring fresh-mixed eggs.

—advertisement, *Restaurant Business,* May 1974

Another breakfast item that has made sensational strides as a frozen item is the liquid, low-cholesterol eggs. . . . However, new claims to scientific tests on humans, indicating that a diet of two eggs a day for young people and one egg a day in older persons produced absolutely no effect on the cholesterol level in their blood streams, could give impetus to liquid egg products in the frozen retail field, where they would represent the ultimate in convenience.

—*Quick Frozen Foods,* December 1975

What I have considered strange in the diet recommendations of the AHA [American Heart Association] is the slight concern towards the consumption of potato chips, french fried potatoes, doughnuts, snack foods and soft drinks. These are all high calorie foods which can be converted to cholesterol in the body. Yet, none of these items have been vilified in the same manner as the egg.

—Dr. Fred A. Kummerow, "The Role of Eggs in the American Diet," Urbana, Illinois: The Burnsides Research Laboratory, University of Illinois, undated

The Southeastern Poultry and Egg Association . . . voted to give its annual recognition award to the McDonald . . . breakfast item, "egg Macmuffin." McDonald, however, was quick to decline. "They told us they didn't want the publicity because of the growing controversy over the role of eggs in heart and arterial problems," a spokesman for the egg producers said. So the award will now go . . . for an item called "egg basket," which doesn't contain eggs.

—*The New York Times,* 31 January 1976

We can congratulate the [food] technologists who devised [substitute products for eggs and egg white]. Nevertheless, underneath our respect for the march of science we may retain a sneaking yearning for an egg preserved merely in its own shell and by the chemical mechanism which it contains in nature.

—Dr. Magnus Pyke, *Townsman's Food* (London: Turnstile Press, 1952)

〉〉〉〉〉〉〉〉

The marketing of substitute egg products was common long before Standard Brands introduced Egg Beaters® to a cholesterol-worried civilization. Fabricated egg products had long been used by bakers and other food processors.

As early as 1938, a German manufacturer began to produce and export Weiking Eiweiss, a product in which fillets of cod or haddock, or dried salted stockfish, or even shrimp, were substituted for the egg white (albumen). The fish was shredded, soaked in a warm solution of acid, and rinsed in water. The fat was extracted from it, either with alcohol or with trichloroethylene (a solvent commonly used for dry cleaning clothes as well as for food use, and now considered a carcinogen), and then the dried, powdered fish was heated in a caustic soda solution long enough to break it down. Then the caustic soda was neutralized with acetic acid and the mixture was spray dried. The resulting product, described as having good color and " a slight, readily masked odor" replaced egg in mayonnaise, French ice cream, many types of bakery products, and numerous other food products. With growing wartime food scarcities, Weiking Eiweiss increased in popularity.

Another German egg-white substitute, made from surplus skim milk, was known as Milei-W, the W being for *weiss,* "white." After extensive processing, the resulting powder simulated egg white, with a foaming property, when added to bakery goods and other food products. The same company also made use of surplus skim milk for Milei-G, the G being for *gelb,* "yellow," a simulated egg-yolk powder, which was used mainly by German food processors of dumplings and batter puddings. This enterprising company then produced a whole-egg substitute, Milei-V, the V for *vollei,* "full egg." By modifying the formula of the earlier versions, adding a

35

yellow dye, and including a specially fermented whey, the whole-egg substitute was—if not exactly hatched—at least devised. German technologists continued in their quest for substitute eggs. One egg-white substitute called Plenora was made, strangely enough, from slaughterhouse blood. The plasma, a clear, straw-colored fluid, was spray dried, mixed with starch and locust kernel gum, and sold to bakers and other food processors. Plenora approximated the equivalent whipping qualities of real egg white and was used to replace whole egg yolk in products such as mayonnaise.

For many years, bakers and other food processors have used substitute egg products extensively. These low-cost products are substituted in part or totally for whole eggs in baking formulas in a wide range of products. The replacers do not require refrigeration and are less costly than whole, fresh eggs. Some of the substitute egg products consist of wheat starch, lecithin, dried egg white, lactose, and synthetic color. Such products simulate the emulsifying and structure-building qualities of dried egg yolk. They are favored especially for use in doughnut formulas and mixes. Their desirable properties allow doughnuts to break normally and not lose volume, and "the fat holdout is comparable to dried egg yolk."

From the viewpoint of food technologists, substitute egg products were a technical achievement that contributed to the "palatability and attractiveness of the diet." However, as Dr. Magnus Pyke noted, "the ingenious substitutes for egg and for egg albumen do not claim the vitamins, the iron, and the rest which eggs themselves contribute to the chemical composition of the diet." The end products made from these animal proteins of fish, milk, and blood were termed by Pyke "nutritionally unexceptional."

The launching of substitute egg products for the retail market was made at a propitious time, with the public alarmed about cholesterol. To many, the cholesterol-free egg substitute seemed like the happy solution. Advertising copy for the new product gave the message, loud and clear:

> Now you can give up the cholesterol, without giving up the honest-to-goodness taste of eggs. With new Fleischmann's® Egg Beaters™. The world's first fresh-frozen, cholesterol-free egg substitute. (And

not just a powdered dehydrated excuse for eggs.) Why is this important to you? Because the Inter-Society Commission for Heart Disease Resources recommends limiting dietary cholesterol to less than 300 mg. a day. The average large egg contains 275 mg. of cholesterol. It's the single highest source of cholesterol in man's diet. Like eggs, Egg Beaters™ give you important vitamins, minerals, and protein. You can scramble them, make delicious omelets, French toast, pancakes, even "egg" salad. And they look, cook and taste like farm-fresh whole eggs. Only no cholesterol. Just defrost Egg Beaters™, and use them like fresh eggs. . . .

How accurate were these claims? What about the honest-to-goodness taste? One taste tester's assessment of scrambled Egg Beaters® was "utterly bland, like eating soft, hot cotton." Others judged the scrambled eggs as "wet, runny, and lacking in egg aroma." "Egg" salad, made from a recipe supplied with the product, was described as having "the taste and texture of chunks of wet cardboard, heavy on the mayo." Omelets "wept" and "lacked aroma." When disguised in meatballs, soufflés, and French toast, where stronger flavors can mask the product, Egg Beaters™ fared better. "I can get away with them in potato pancakes," admitted one tester. Because of the low-fat content of the product, batters were found not to "set" properly, so cakes came out with a heavy, grainy texture.

How accurate was the claim that the egg substitute "offers the nutrition of fresh eggs"? One large whole shell egg contains about 7 grams of high-quality protein. Though the label claims for the egg substitute a protein content of 6.6 grams for each two-ounce serving, laboratory analyses of samples showed it to be only 5.4 grams, which is 18 percent less protein than claimed and 23 percent less than in one whole egg. (Eggs, which have *all* essential amino acids, contain them in such a well-balanced proportion that they approach the theoretical ideal protein for humans. For that reason, eggs are used as the reference standard in evaluating the protein of all foods.)

The substitute egg product label states that four vitamins and five minerals have been added, which may appear impressive, but the additions do not make the product nutritionally equivalent to whole shell egg. A real egg yolk contains significant amounts of two minerals essential for good human nutrition, phosphorus and manganese. The ingredients used to formulate the egg substitute product lack significant amounts of these nutrients. The product lacks all usable sulfur, Vitamin A, and niacin, and almost all zinc, as well as other trace minerals. The substitute egg product is fortified with two B vitamins (thiamine and riboflavin), whereas a whole egg contains not only significant amounts of both but also pyridoxine and pantothenic acid, contained in the yolk. Although the egg substitute product may contain some pyridoxine and pantothenic acid, they are not present at the same high level, nor in the same ideal balance.

Analysis showed that the substitute egg product has about ten times as much carbohydrate (sugar and starches) as real egg. The high carbohydrate content may be undesirable, especially for those who are obese, on weight-reduction diets, or diabetic.

The egg substitute product is higher in sodium than a real egg. Tests showed that two ounces of the egg substitute averaged 140 milligrams of sodium, or 28 percent more than advertised, and 133 percent more than contained in one large whole shell egg. This sodium content is especially undesirable for those on low-sodium diets.

Since heart patients are frequently placed on weight-reduction, low-cholesterol, or low-sodium diets, it becomes apparent that the substitute egg product may not serve the needs of these individuals.

One would assume, before any food product is successfully devised, launched, and accepted by a segment of the consuming public, that it has undergone thorough testing for its nutritional qualities. This would be especially true for any product with special features that alter or modify the traditional diet.

However, the question, "How well do the egg substitute products meet growth requirements?" was raised, oddly enough, only *after* the products had been marketed.

A comparison was made of the nutrients in equal amounts of cholesterol-free egg substitutes and farm-fresh eggs. An examination of the list of nutrients contained in the substitute egg products made it appear that such products should meet the growth requirements of weaning rats. However, experiments showed that they did not. M. K. Navidi and F. A. Kummerow found that pups from the mother rats fed the egg substitute products averaged only 31.6 grams in weight at three weeks of age; those fed whole eggs achieved 66.5 grams, or more than double. Animals fed a stock diet did slightly better, with a weight of 70 grams. Both the mothers and pups fed the substitute egg products developed diarrhea within a week, whereas those fed whole egg did not. The pups fed the egg mixtures were weaned at five weeks of age. All those fed the egg substitutes died within three to four weeks after weaning. The general appearance of those rats indicated "a gross deficiency in one or more nutritional factors" as compared to those fed whole eggs. Navidi and Kummerow concluded: "It is evident that shell eggs, which contain . . . egg yolk, furnish one or more nutritional factors which are absent in Egg Beaters®. These nutritional factors are no doubt present in the common food items which comprise the diet of human adults. . . . [but] may not be present in adequate amounts for infants fed milk and Egg Beaters® instead of egg yolk from a soft boiled egg."

Navidi and Kummerow concurred with the recommendations of the Council of the American Medical Association, which had stated, "Care [must] be taken to assure that the dietary advice given does not compromise the intake of essential nutrients." The council had recommended that no drastic dietary changes be made: it was considered unwise to remove cholesterol from infant foods, curtail the amount of eggs, meat, and dairy products consumed by growing children, replace such foods with polyunsaturated oils, and so forth. Any such radical dietary changes might, in the opinion of the council, "result in nutritional disaster."

5. Old Bossie's Cream versus Imitation Cream Products

《《《《《《《《《

In general, homemakers identified the inviting qualities of imitation dairy products as price, calorie count, convenience, and keeping characteristics. On the other hand, they appraised genuine dairy items as superior to imitations in taste, food value, and purity or absence of harmful additives.

—"Homemakers' Opinions About Dairy Products and Imitations: A Nationwide Survey," *Marketing Research Report No. 995,* Statistical Reporting Service (Washington, D.C.: USDA, May 1973)

It was some comfort to learn . . . that I was not alone in my war against the synthetic creamer. My battle tactic has been to signal the [airline] stewardess, preferably at her busiest moment, and request sweetly but loudly for some real milk (I know it is hopeless to ask for cream) because, as I inform her, "I won't have anything to do with this awful junk." . . . The most discouraging part is that I am invariably the only passenger who objects. Everyone else docilely accepts the substitute.

—Paul J. C. Friedlander, "That Ersatz Cream," *The New York Times,* 27 February 1972

Imitation milk is now being marketed by two [conglomerates] and is expected to take up to ten percent of the fluid milk market by 1980. Pet, Car-

Old Bossie's Cream versus Imitation Cream Products

nation and Borden are brand names that once were synonymous with real milk and milk products, but today these firms are among the largest marketers of such artificial milk products as non-dairy coffee "whiteners." Even the Department of Agriculture's main cafeteria in Washington [D.C.] recently has taken the symbolically significant step of switching from coffee milk to whiteners.

—Jim Hightower, "Fake Food Is the Future," *The Progressive,*
September 1975

Nutritionally balanced milk substitutes: A milk analogue which can be produced in existing spray-drying facilities, with slight modifications, and based on wheat flour, soybean oil and skim milk, has been developed for preschool and school-age children. In tests with 6,500 children, the lower-cost substitute product available in chocolate, coconut, banana, vanilla and strawberry flavors rated higher in acceptability (95 percent versus 75 percent) than milk used in the feeding program.

—*Food Processing,* January 1975

Synthetic creams and meringues are today often made from cellulose derivatives which have absolutely no nutritional value. . . . The more luxurious classes of foodstuffs lend themselves particularly well to sophistication and it may be argued that these are not, in any case, eaten primarily for their nutritive value. This is a specious argument, however, and does not wholly remove the impression that the use of such substances is tantamount to a confidence trick.

—Brian A. Fox and Allan G. Cameron, *A Chemical Approach to Food and Nutrition* (London: University of London Press, 1961)

Freeze and Thaw Imitation Sour Cream, U.S. Patent 3,729,322: A food product having the flavor and consistency of sour cream which is capable of being frozen for an extended period of time and subsequently thawed while still retaining its smooth texture and which contains a unique combination of hydrophilic colloids including cross-linked amioca starch, low methoxyl pectin and hydrooxypropyl methyl cellulose.

—*DRINC* (Dairy Research, Inc.), June 1973

The Great Nutrition Robbery

To assist consumers who may be confused by certain imitation products . . . the California Milk Advisory Board has designated and trademarked an identification symbol for genuine dairy products. . . . Our research tells us that many people purchase imitation dairy products not knowing that they are ersatz, chemical substitutes for the real thing. Clever packaging that looks like genuine dairy product cartons . . . probably confuses many shoppers.

—*American Dairy Review,* October 1977

Things are seldom what they seem.
Skim milk masquerades as cream.

—*Gilbert and Sullivan, H.M.S. Pinafore*

〉〉〉〉〉〉〉〉

Chemistry has been replacing old Bossie. By 1974, one fourth of the entire dairy market was estimated to have been lost to nondairy imitators within a twenty-five-year period. Factors in the trend include greater convenience, longer shelf life, and lower cost of the nondairy products. The growing interest in low-calorie, low-fat, low-

USDA STUDY, 1970

PRODUCT	PACKAGE UNIT	AVERAGE ADVERTISED RETAIL PRICE (IN CENTS)	ESTIMATED INGREDIENT COST (IN CENTS)
Whole milk	½ gallon	50.8	29.1
Filled milk	½ gallon	37.4	20.3
Coffee cream	pint	43.3	19.6
Nondairy creamer	pint	20.2	5.4
Whipping cream	pint	49.9	29.4
Nondairy whipped topping	pint	31.6	4.6
Sour cream	pint	66.8	21.4
Imitation sour cream	pint	45.6	6.9
Ice cream	½ gallon	67.2	25.0
Imitation ice cream	½ gallon	41.4	13.3

cholesterol food has stimulated increased use of nondairy products despite the fact that such items are not necessarily low calorie or low fat. Also, the type of fat they contain is usually highly saturated, even more than butterfat, which it replaces.

Cost is definitely a factor, as demonstrated in a study made by the USDA in 1970. The average advertised prices and estimated ingredient costs for selected dairy products and their substitutes were compared.

Margarine (see Chapter 7) was the first in a long list of simulated dairy products. It has been available for more than a century. More recently, many types of nondairy products have been developed that replace real milk and cream. The list continues to grow:

DAIRY PRODUCT	REPLACED BY
Creamery butter	Margarine
Coffee cream	Nondairy creamer
Whipping cream	Nondairy whipped topping
Fluid cow's milk	Filled milk and imitation milk
Low-fat milk	Imitation low-fat milk
Milk shake	Nondairy shake
Ice cream	Mellorine
Sour cream	Imitation sour cream
Buttermilk	Imitation buttermilk
Evaporated and condensed milk	Imitation milk concentrates
Dairy snack dips	Nondairy snack dips
Cheese	Imitation cheese

Segments of the dairy industry, instead of viewing these developments glumly, have decided to produce both nondairy items and traditional ones. Robert Anderson, executive director of the National Cheese Institute, urged the dairy industry to be open-minded about imitation products: "We must try to make them in a manner so they won't detract from natural products. If we don't do it, others will, and they may do more harm."

Coffee creamers hit the market in 1952. A company producing an infant feeding formula found that it needed to make economical use of a by-product, cream, since the formula made use of the noncream constituents in milk. A midwestern company began marketing the cream in powdered form as a coffee creamer. Although the new product was made from cream, Americans were used to the taste of fresh cream and milk. The new product had a characteristic taste, somewhat like evaporated milk. The product went back to the drawing board, or rather to the research laboratory. By substituting coconut oil (selling then at eighteen cents a pound) for butter (at seventy cents a pound), a modified product was formulated that, according to the researchers, "tasted better, dissolved quicker, had greater stability, and cost less."

However, the new product was illegal under the dairy laws of most states, so the word *creamer* was dropped and replaced by the term *nondairy creamer*. That action was considered daring, since it was thought that consumers would not be attracted to the idea of a term that was negative and implied that the product was synthetic.

But the new products were launched at the time of the cholesterol scare. Although the fat level of the coffee creamers and nondairy creamers was basically the same, people inferred wrongly that the new products were low calorie, low fat, and low cholesterol. The new products gained in popularity.

The dairy interests fought back. They wanted the new products to be labeled imitation cream so that consumers would not be confused. Producers of the new products stoutly maintained that nondairy creamers were not imitation cream, which implies inferiority, but rather something new that required classification in a new product category. One company won court cases in every state that challenged the right to sell nondairy creamers. By 1976 the cases were won in thirty-nine states. The most important decision was rendered in Wisconsin, a strong dairy state, where serving nondairy creamers in restaurants had been prohibited. When the statute was challenged, the Wisconsin Supreme Court unanimously declared it illegal. The court held that the nondairy creamer is "a wholesome, nutritious *sui generis* [i.e., constituting a class alone] food product and not an imitation" and that the product "enjoys advantages over cow's cream in its resistance to spoilage, price and stability."

The Wisconsin statute was considered unconstitutional as a burden on interstate commerce. By this time, the easy-to-open, convenient one-portion packs of nondairy creamers were moving across state lines everywhere and being delivered to restaurants, institutions, and airplanes. They largely replaced the cream pitcher and individual serving jugs of coffee cream. Such jugs have become collectors' items.

Some 3.5 to 4 billion half-ounce-portion nondairy creamers are served to Americans each year. That figure may reach as high as 11 billion in the near future. Although some restaurants that still take pride in serving quality food may serve cream or "half and half," many restaurants as well as hotels, motels, diners, and other places where food is served have come to favor the use of nondairy creamers.

High-speed machines fill, seal, and pack the creamers quickly and efficiently. The packaging system minimizes waste through spoilage or spillage. However, such packaging requires high energy use and the products make extravagant use of nonrecyclable plastics and heat-seal-coated film lids.

For institutional use, including restaurants, nondairy creamers are packaged in pint, quart, and half-gallon cartons. These products, delivered frozen, are thawed out as needed. After thawing, the product "stays perfectly fresh for three weeks under normal refrigeration; up to three hours on the table." That is the promise made to restaurateurs in advertisements for these products in trade journals. The products "whiten better so your customers don't use as much," and the products "taste as rich, whiten as well as half-and-half," but "cost up to 50 percent less." One product was described as "rich enough to be used over fruit and desserts."

In addition, nondairy creamers, liquid and dry whipped toppings, and imitation creams are now widely available in retail food stores for home use. They have become popular because of their convenience, long shelf life, and price. The nondairy creamers are being used to replace cream or milk with cereals and fruits, or as cooking ingredients in recipes calling for cream or milk, as well as in coffee or tea. It is estimated that 40 percent of the dairy cream market had been lost to liquid and dry nondairy creamers by the mid-1970s.

What do nondairy creamers consist of? Even the conscientious label-reader has difficulty deciphering the labels on the individual portion packs of these products. Frequently, the almost illegible list of the ingredients is printed in minuscule type in one, and at times even two, circles around the circular cover. To many people, the listing appears to be merely part of the ornamentation. The consumer is further confused since real dairy products, such as half-and-half, are packaged in a form identical to that of nondairy creamers. Some dairies also process nondairy creamers, so the name of a dairy on the label makes it appear that the product is from a dairy source.

Food laws and regulations require a generic term as a basic part of the marking of all food products. Nevertheless, vague and at times fanciful terms are tolerated by food control officials. Examples include names such as "For Your Coffee," "Coffee Dream," "Coffee Twin," "Coffee Pal," "Coffee Companion," "Coffee Rich," "Coffee High," "Instant Creamer," "Café Creamer," "Coffee Creme," "Creem'r," and "Cup-A-Creme." Of course, "creamer" and "creme" should not be confused with real cream.

Nondairy creamers include an astonishing array of ingredients, mainly chemicals. In addition to water, some hydrogenated oil is included, which usually consists of coconut oil or palm kernel oil. Less frequently, soybean oil is used. At times, the vague terms *vegetable fat* or *edible oil* give no inkling of the specific substance used. Sodium caseinate solids (from milk) or, less frequently, "vegetable protein" (probably from soybean) are included. Sweetening agents generally consist of sucrose (sugar), corn syrup solids, and, less often, lactose (from milk) or sorbitol. Then come a number of chemical additives on the ingredient listing, including sodium or calcium phosphate or di-potassium, sodium citrate; propylene glycol monostearate, polysorbate 60, mono- and di-glycerides, sorbitan monostearate, or sodium stearoyl 2-lactylate; carrageen and guar gum; sodium silico aluminate; salt; lecithin; artificial color and flavor, and at times a preservative, benzoate of soda! Some of these additives are objectionable for health reasons. Use of so many additives also raises a basic question as to why it is necessary to go to such an extraordinary degree of sophistication for a cream substitute.

The term *nondairy* is a misnomer, since most of these products do contain sodium caseinate (a milk protein component that is judged to be less valuable than other milk proteins), and a few of these products contain lactose, a milk sugar. For the large number of persons who are highly sensitive to milk products caused by lactose in- tolerance or milk allergy and must avoid all constituents of milk, the term *nondairy* is misleading.

At one time, and for reasons known only to the FDA and possibly to the nondairy product manufacturers, the FDA ruled that sodium caseinate is a chemical product, not a milk product. So the term *nondairy creamers* may be within the law, though clearly it does not make for honest labeling for those who need to avoid milk compo- nents. A few companies label their products more correctly by stat- ing on the label of the products that they contain no milk fat.

Generally the fat used in nondairy creamers is coconut oil or palm kernel oil. Both are highly saturated. Some individuals choose non- dairy creamers with the mistaken notion that they are reducing their caloric intake with a low-fat product, but this is not the case. Non- dairy creamers have at least as many calories as whole milk and they contain *more* total saturated fat than is found in the butterfat of milk.

For diabetics or other individuals who need to avoid sugar, it is important to note that all nondairy creamers contain sugar in some form. For persons on sodium-restricted diets, it is important to note that many of these products contain salt, as well as sodium-contain- ing additives.

The extraordinarily long list of chemical additives used with non- dairy creamers allows these products to achieve an unusually high degree of stability. Such products must remain in a uniform and ac- ceptable physical state long after their manufacture and sale to the store or restaurant. Their formulas are such that these creamers disperse well when added to hot liquids such as coffee or tea. Some of the chemicals used in these products control viscosity in order to simulate as much as possible the texture and flow characteristics of real cream or milk. Some components in these products provide uniform whitening ability, which is determined by the total amount of solids they contain.

Nondairy creamers have been described by Dr. Jean Mayer as a "formula for coloring coffee white, and nothing more." He singled out nondairy creamers as a typical example of a synthetic food that competed with a natural one and lacked nutritional value. "By far, the biggest culprit against healthy eating is our own free choice" with food selection. "You may argue, one tablespoon [of a nondairy creamer] makes little difference, one way or the other. But in a nation where overweight is common and is a risk factor in some of our most prevalent causes of death and disability . . . we should make every calorie count in terms of nutrients. And few of us stop at one cup of coffee. The calories mount up with each tablespoon; if it's milk that's being added, you also get nutrients." Mayer concluded that there may be a place for nondairy creamers, such as for airline food service where milk is difficult to store and serve, but "there should be no place for it in your family kitchen."

Imitation Milk

Like nondairy creamers, "filled" milk and "imitation" milk are inexpensive substitutes for real dairy products. For example, the ingredients in imitation milk made from soluble soy protein isolate and other raw materials may cost approximately 71 percent less than cow's milk, but the savings are not passed along to consumers.

Filled milk is defined as "a combination of any milk, cream or skimmed milk with any fat or oil other than milk fat, so that the resulting product is an imitation or semblance of milk, cream or skimmed milk." A filled milk product that does not meet the FDA's definition of "nutritional equivalency" of the product it resembles is considered to be an imitation.

For a filled milk product to be considered nutritionally equivalent to its counterpart milk product, it must contain a specified quantity of minimum nonfat milk solids and Vitamins A, D, and E. The FDA recognizes that some vegetable fats are characteristically deficient in certain polyunsaturated fatty acids, so the agency has ruled that polyunsaturated fatty acids of the linoleic series must be present at a level equivalent to 4 percent of the fat in the filled product. Interest in formulating filled milk and imitation milk products has coincided with the cholesterol scare.

To date no clinical studies have been conducted to substantiate the claim that filled milk products that comply with the standards defined by the FDA for nutritional equivalency are, in fact, nutritionally equal to their counterpart milk products. On the contrary, one study demonstrated that cow's milk has four to nine times the amount of calcium found in imitation milks and about three to four times as much phosphorus. Also, *nutritional equivalency cannot be measured solely on the basis of totaling quantitative nutrients.* At present, little information exists regarding the important characteristic of biological availability of the components in filled and imitation milk products, nor in many other fabricated foods.

What is known is that the fat portion of the filled milk products is distinctly different from its conventional dairy counterparts. Milk fat has a unique and complex physical, chemical, and biological profile. It is virtually impossible to substitute for this milk fat an exact equivalent fat or blend of fats. Nutrient interactions with various fats are quite distinctive and result in clearly defined differences in products. Despite these differences, the FDA's proposed filled milk regulations failed to establish any fat specifications in filled milk products. As with nondairy creamers, the fat used in many of the filled milk products is highly saturated and hydrogenated coconut oil.

In a comparison of the relative nutritional value of filled and imitation milks with cow's milk, one significant feature of the fabricated milks turned out to be their diversity. As a group, they cannot be uniformly depended upon to supply certain nutrients. Some were devoid of Vitamin A, whereas others contained an amount that might be considered excessive and toxic. Imitation milks were found to be devoid of riboflavin, low in total protein and essential amino acids, and low in both calcium and phosphorus. Investigators unanimously agreed that "the imitation milks and certain filled milks as formulated today are unsuitable for infants and children." The products were judged to be "potentially harmful for other vulnerable age groups such as pregnant and lactating women and persons on marginal diets such as those in low-income groups and the aged." These products are also considered to be unsuitable for persons on fat-modified diets because of the highly saturated coconut oil content.

The National Dairy Council, concerned about the increased marketplace importance of filled and imitation milks replacing cow's milk, had an independent laboratory perform a comparison of nutrients in selected samples. Results were made available in 1968, with values of gram/100 gram:

FILLED MILK

NUTRIENTS	WHOLE MILK	SAMPLE #1	SAMPLE #2	SAMPLE #3	SAMPLE #4	IMITATION MILK
Total fat	3.420	3.775	3.125	3.365	3.485	3.685
Total saturates	71.495	97.810	92.990	90.620	87.985	94.835
Total unsaturates	28.505	2.19	7.010	9.380	12.015	5.170
Total nonfat solids	8.885	10.215	9.130	8.840	9.415	8.085
Total protein	3.380	4.085	3.365	3.080	3.060	0.880
Total ash	710.00	865.00	830.00	735.00	750.00	505.00
Calcium (mg/100 gm)	125.50	148.50	143.00	117.00	101.00	23.85
Phosphorus (mg/100 gm)	99.50	114.50	102.50	94.50	71.00	14.25
Vitamins: Total Vitamin A value (USP units/qt)	447.00	none	1195.00	none	none	4010.00
Riboflavin (mg/qt)	0.725	0.790	0.860	0.755	0.820	none
Total carbohydrate	4.795	5.265	4.935	5.025	5.605	6.700

Nondairy Whipped Toppings and "Cream"

A frozen nondairy whipped topping, intended as a substitute for whipped cream, was produced and marketed in the United States as early as 1945. The product, made from soybean, was packaged in a cone-shaped container similar to a type used at that time for real cream.

At the beginning, most of the sales of the product were at the retail level. In time, institutional, business, and bulk sales of the product grew. The nondairy whipped topping stored almost indefinitely in frozen form. Once thawed, it whipped up to triple its original bulk in forty-five seconds. These features impressed food processors. Furthermore, trade journal advertisements informed processors that the nondairy whipped topping allowed for an attractive margin of profit.

The early success of frozen cream pies was due almost entirely to the development of a coconut oil–based nondairy product, judged to have excellent flavor and texture. The new product had none of the bacterial problems associated with real whipped cream. Such non-dairy products were especially favored in areas of the country where extreme heat creates problems with highly perishable foods such as cream.

Another synthetic cream was created by combining about 15 percent sodium alginate and 5 percent methyl ethyl cellulose. The product had very satisfactory whipping properties.

Since whipped topping competes with cream, it was thought that such nondairy products would do less well in states with strong dairy interests. But this has not happened. Nondairy whipped toppings sell well in states like New York, an important dairy state, probably as a result of price, convenience, and shelf life. By 1974, nondairy whipped toppings controlled about 85 percent of the whipped topping market.

Restaurants and other institutions make extensive use of nondairy whipped toppings. The products are available in pressurized cans, are frozen in milklike cartons, and can be bought in powder form. Restaurateurs are informed that the products from pressurized cans are "the smoothest, richest tasting whipped topping you ever squirted." The product "quickly makes something special out of pies, cakes, sundaes, gelatins, or any dessert" and "holds up much longer, much better, even on problem desserts like hot items." Frozen whipped toppings are offered as being "smooth and stable, so you can prepare toppings hours ahead." Processors of the toppings in powder form assert that "for centuries there was no substitute for the function of dairy fresh cream in fine foods, desserts and coffee" until this manufacturer, in developing its product, "did nature one better" with "wonderful whippable nondairy delights" and that the product "can be rewhipped and rewhipped."

Uses for these products continue to increase in number, with product innovation. One large restaurant chain announced that new recipes and concoctions were constantly being developed for soft-serve ice cream. "One of the interesting ideas we've tested recently is a non-dairy powder mix which is being developed especially for [us] by a commercial laboratory. The vanilla was well received but the chocolate still needs some work. . . ."

On the retail market, chocolate and artificially flavored strawberry nondairy whipped toppings have been introduced. Suggested uses for these new products "are limitless—to top fruits, cakes, pies, sundaes, and cookies; to mix in malts and floats; to fill tortes and éclairs; even to float in coffee or blend into gelatin."

Nutritionally, nondairy whipped toppings have little to commend them. A typical formulation consists of about 30 percent fat (of the saturated type) to provide the richness, body, and texture characteristics of whipping cream; an undesirably high 10 percent sugars (from sucrose and corn syrup solids); a small 2 percent protein (from sodium caseinate); and a high percentage of water, with an emulsifier, stabilizer, stabilizing salts, flavor, and color.

Imitation Sour Cream

Imitation sour cream, in regular, powdered, and canned forms, is usually formulated from vegetable fat, protein (sodium caseinate or nonfat milk solids) mono- and di-glycerides, buffer salts (phosphates), modified starches or gums, citric or lactic acid, and flavoring. Mixtures of these ingredients may be blended, homogenized, then spray dried to produce a powder that can be reconstituted with water.

Mellorine, An Ice Cream Substitute

In mellorine, vegetable and animal fats replace butterfat. Although mellorine contains some nonfat milk solids, its main ingredients are nondairy; mellorine is not considered a dairy product. Parevine, though similar to mellorine, contains no animal fats.

Mellorine has replaced about 15 percent of the frozen dessert market and by the mid-1970s had an annual sale of 3.2 million gallons. This figure would be higher except that the product is legal in only about one fourth of all states.

Mellorine consists of about 20 percent nonfat milk solids, from 15 to 16 percent sweeteners, and 8 to 10 percent fats, plus a stabilizer, emulsifier, and other additives commonly used in frozen desserts. The vegetable fat may be cottonseed oil, which, though nutritious, is undesirable for food use because of current agricultural practices. The cotton crop is not classified as a food crop, so it receives more pesticides than any other crop. High levels of pesticide residue contaminate the plants and are especially concentrated in the oil. Other fats used in mellorine include tallow from meat and oils from coconut, corn, peanut, and soy. Both the meat fats and coconut oil are highly saturated.

Technology Gone Mad?

Concerning the fabrication of one nondairy product, Dr. Jean Mayer commented, in a statement applicable to all nondairy products, "What I fear is technology, in a sense gone mad, producing pseudo foods which will fool busy or inattentive people into thinking that they are consuming nutritionally valuable foods when, in fact, they are consuming calories that make no contribution whatsoever to human nutrition. . . . Let us make sure . . . that with an ever-mounting number of pseudo foods we don't get into a situation where the benefits are the manufacturers' alone while the consumer has all the risk."

6. Imitation Cheeses Replace Natural Cheeses

〈〈〈〈〈〈〈〈〈〈

U.S. Patent 3,741,774. A high protein simulated cheese product is prepared by forming a mixture of specified amounts of cheese, pre-gelatinized starch, a high protein binding agent, water and sugar or sugar equivalents. The mixture is heated to 125°F. to 195°F. and while at a temperature within this range it is extruded into small strands.

—*DRINC* (Dairy Research, Inc.), August 1973

The flavor of imitation cheese is developed by adding artificial flavor, rather than through a time-consuming and costly aging process. . . . An urgent and . . . immediate problem is to improve the flavor of imitation cheese products. So far, the flavor of natural cheese has not been duplicated. Although improvements in flavor are being made almost daily, the ultimate accomplishment has not been realized.

—*American Dairy Review*, June 1976

What does imitation cheese taste like? Comparisons that come most quickly to mind are kneaded rubber erasers and crayons.

—Mimi Sheraton, "Tasting Test for the Ersatz," *The New York Times*,
2 February 1977

Imitation Cheeses Replace Natural Cheeses

The real reason behind the introduction of imitation cheese is not health, but simple economics. The cost of oils made from cottonseed, corn, coconut, or soy is much less than that of milk fat; imitation cheeses are therefore cheaper to produce. . . . However, while some consumers will find that some analog cheeses are cheaper, experts believe that eventually the analogs will cost shoppers as much as real cheese.

—Rona Cherry, "Forget Cheese, Smile and Say 'Analog,' "
The New York Times, 2 February 1977

An extensive line of cheese "enhancers" can be used for such products as cheese sauces, analogs and snack foods. The flavors, which include artificial Cheddar cheese, artificial Parmesan cheese and artificial American cheese, can be substituted for up to 50 percent of the natural cheese solids. Also offered is artificial Mozzarella cheese flavors, used to enhance the flavor of Mozzarella cheese.

—*Canner/Packer*, January 1976

Imitation cream cheese emulsion, an authentic reproduction of the flavor of cream cheese, permits development of a variety of brand new products and improvements of old favorites. Effective in imparting a mild dairy note to cakes, sweet doughs, doughnuts, butter creams, and icings, [the] emulsion never varies in strength or quality. Unlike culture-derived flavors, it is stable at room temperature and does not require refrigeration.

—*Baking Industry*, December 1975

Cholesterol-free imitation cheese: made with cotton seed and sunflower seed oils, and skim milk, cholesterol-free imitation cheese is recommended for individuals modifying their total dietary intake of cholesterol . . . imported from Sweden . . . [the] cheese is available at deli counters and cheese shops.

—*Food Processing*, May 1976

U.S. Patent 3,946,123. A canned pet food consisting essentially of an aqueous medium of 0.05 to 15 weight percent of a flavoring agent selected from the class consisting of cheese, cheese flavoring agents.

—*DRINC* (Dairy Research, Inc.), May 1976

》》》》》》》

55

Imitation Cheese

Imitation cheese is not a recent innovation. Prior to 1962 in Italy, where citizens were shocked by a series of incidents in which the skills of unethical food "sophisticators" were exposed, imitation cheese played a pivotal role. Until then Italy had no federal police force to combat food frauds. Through the years, as each fresh food scandal broke, Italians sat down to dine, either amused or upset, but not surprised.

In October 1962, Colonel Francesco Naso, attached to the Roman Legion, read a newspaper account of the latest outrage, the creation of "cheese" from ground-up buttons and umbrella handles. Naso, described as an emotional man driven by honor, promptly sought official permission to organize a special police force to fight food frauds.

The trail of the cheese fraud led to a routine check of a cheese maker who reported to Naso that he usually fed the cheese residue to his pigs. However, some Milanese cheese packers had bought his residue for a goodly sum and claimed that they would use it to breed trout.

Naso discovered that the Milanese had no trout. They were mixing the cheese residues, which possess concentrations of both the odor and the taste of cheese, with a mixture of oils from fish, soybean, coconut, palm, and margarine to create a product that had the taste, smell, and consistency of Parmesan cheese. Its lack of genuine yellow coloring was corrected by adding carotene and saffron. This product was sold as genuine Parmesan cheese until Naso and his cadre raided.

Another "cheese" Naso discovered was made from banana peels. This waste is rich in sugar, proteins, and starch, as well as in cellulose and pectin needed for the proper consistency.

Imitation Cheese Reduces Real Cheese as an Ingredient

By the mid-1970s in America, the high cost of food, including cheese, made the time propitious to develop nondairy cheese substitutes. The price of a forty-pound block of cheddar cheese—a cheese that essentially determines the price of all other cheeses—had increased by 40 percent from 1975 to 1976.

Many factors were operating to affect the price of cheese. Import cheese quotas were restrictive, and domestic milk production declined. The number of cheese plants decreased. From 1960 through 1973, the number of cheese plants tumbled from 1,419 to only 865. At the same time, consumption was increasing, both for table cheeses and for cheeses used in processed foods. Between 1960 and 1974, consumption of natural cheese jumped a whopping 96 percent, with almost half the increase after 1971. Italian-type cheeses now account for more than 20 percent of all natural-cheese production, possibly because of the increased popularity of pizza. The annual consumption of mozzarella rose from about 0.2 pounds per person in 1957 to 2.0 pounds per person in 1973. All these factors combined to increase the price of cheese and create shortages. The extraordinary demand for natural cheeses opened the way for the imitations.

Imitation cheeses and cheese products are usually made of vegetable oils (frequently highly saturated coconut oil), protein, hydrolyzed cereal solids, buffer salts, color, and flavor. The texture, viscosity, and mouthfeel of soft, processed cheeses and spreads, as well as hard cheeses and grated cheeses, have all been simulated.

Casein, the milk protein found in natural cheese, would be the logical choice as the protein source in an imitation cheese. But since casein is insoluble, it offers some technological problems. Some solubilizing agent is needed to make the casein usable in imitation cheese production. To avoid the problems, many manufacturers choose to use some compound of casein. Calcium caseinate is preferred, because of the need for a calcium level in imitation cheese that approximates that in natural cheese, but sodium or potassium caseinate may be added to increase water absorption in the finished product. The choice of caseinates is determined by function, flavor, and—importantly—price.

Initially, the food industry had mixed reactions to imitation cheese. Flavor was a major area of debate. But while the flavor debate continued, the sales of imitation cheese products climbed.

57

The outstanding advantage of the nondairy cheeses was their lower ingredient costs. Imitation mozzarella and American cheeses cost only about two-thirds as much as their natural counterparts. The price difference between imitation and natural cheeses is expected to be even greater in the future, as costs of raw milk used in natural cheeses continue to climb. Other advantages of the imitation cheeses include an assurance of stable sources of supply and elimination of the problem of seasonal variation.

Three groups of consumers that traditionally use large quantities of cheese have been particularly receptive to imitation cheeses. One is the popular and rapidly expanding pizza industry. Another is the federally supported school lunch program, for which cheese is accepted as a protein portion of the meal. The third consists of various food processors who manufacture foods that contain cheese. All three groups have discovered that imitation and natural cheeses can be combined and further "enhanced" with artificial cheese flavors, to develop at reduced costs the characteristics they desire in their products. In some instances, the imitation cheeses are suggested as total replacements for real cheese.

Advertisements for imitation cheeses and imitation cheese flavors, aimed at food processors, claim that these substances can replace up to two thirds of the more costly real cheese solids they would use in their products. One company states, "These pretenders will maintain quality by fortifying the flavor of cheese solids and hold production costs by reducing the amount of cheese solids needed. All of the cheese pretenders provide savings of from 20 to 40 percent of present cheese ingredient costs." Cheddar, Swiss, bleu, Parmesan, and Romano flavors are offered for use in snacks and in hot and cold products such as sauces, bakery goods, and dips. Imitation cheeses are suggested as total replacers for use in sandwich spreads, enchiladas, tacos, cheeseburgers, and similar products.

Some of the imitation cheeses are used for their good handling properties, which permit thin slicing or shredding at high speed with no sticking together or gumming of the blades. Others are sold in little cheeselike nuggets or as crumbles, cylinders, and gratings and are suggested for blending into soups, salads, dressings, bakery goods, snacks, and hors d'oeuvres. They are billed as "a dramatic new food discovery with the taste and color of real cheese, lower in cholesterol," and for use with canned entrees where the nuggets "won't melt when heated (like real cheese does)." The product is reported to have a shelf life, unrefrigerated, of up to a year. Mozzarella replacer has a refrigerated shelf life of approximately a year, compared to natural mozzarella which, under refrigeration, begins to deteriorate after approximately three months.

Lack of mold growth on the replacer cheeses during storage is an asset for processors. For consumers, the speed with which perishable foods mold may serve as an indicator of their high biological quality. An eminent nutritional researcher, Dr. Elmer V. McCollum, advised, "Eat *only* foods that rot, spoil or decay—but eat them before they do!"

Cheese Flavorings Permit Further Reduction of Real Cheese as an Ingredient

Artificial cheese flavors (as opposed to imitation cheese), in paste, spray-dried, and powdered forms, have been designed as adjuncts to compensate for up to 50 percent of the reduction of the natural cheese solids used in products such as cheese sauces, analogs, and snack foods. Some artificial cheese flavors devised by one company for the wholesale trade include a Parmesan type, a cream cheese type, bleu cheese type, mozzarella type, and mild and sharp cheddar types. These flavors are recommended for use with baked foods, processed cheeses, sauces, dips, and gravies.

Another company offers new artificial cheese flavors "comparable in taste and with similar compositions to natural cheeses" for English and German cheddars, Roquefort, Swiss, Tilsit, and Provolone, designed "to improve the taste of countless cheese-flavored products while cutting production costs by reducing the proportion of natural cheese required." The new cheese flavors can replace cheese powders or serve to enhance them. They can also be used to intensify the flavor of processed cheese. The flavors reportedly cut raw material costs and reduce the tendency of products containing cheese to darken or become bitter during cooking processes.

"Natural tasting, artificial cheese flavors give five times the flavor strength of genuine cheeses, but cost less than one half" is the claim made for a line of powdered imitation cheese flavors. "One pound of the flavor is equivalent to 30–35 pounds of natural cheese." The flavors are recommended for use in crackers, extruded snacks, and spreads.

Within the world of artificial cheese flavorings, specialty items have been created for restaurants and institutions. Some have been devised especially for use in such products as salad dressings, canned sauces, gravies, and frozen foods. For example, there is a canned ready-to-serve artificially flavored cheese sauce intended for use as "a casserole base, with vegetables or baked seafoods, in dips or spreads, on potatoes, sandwiches, or eggs." Other products have an artificial smoky cheese flavor described as "capturing the elusive taste of natural cheese" in a convenient, stable, and highly concentrated form.

Under FDA regulations, the word *imitation* now need be used only if the food is nutritionally inferior to a product for which it substitutes. Imitation cheese products, in which nondairy ingredients supplement or replace any or all of the milk nutrients, do *not* meet the Standards of Identity established for cheese or cheese products. But such products need not be labeled imitation unless they fail to meet the FDA's nutritional standards.

Recognizing the need for clarification, in 1975 the National Dairy Council stated that "although cheese analogs may have a nutrient profile similar to that of the natural cheese variety they purport to simulate, there have been no data published regarding their nutritional value as compared to that of natural cheese. To clearly establish their nutritional quality, it is important that these cheese analogs be subjected to biological evaluation, i.e. animal feeding studies."

Long-overdue feeding studies and analyses were conducted by the Wisconsin Alumni Research Foundation and other laboratories. Not all imitation cheese products met the nutritional equivalencies of natural cheeses. Many imitation cheeses on the market failed to meet the specifications of the Food and Nutritional Services of the

USDA or those contained in the USDA Handbook No. 8 (now superseded by No. 456) on food composition.

The extent of the use of artificial cheese flavorings in fabricated foods and whether the labeling of such products is misleading forms the basis for an ongoing FDA survey. The agency has raised the question of whether items such as cheese-flavored popcorn, potato chips, tortilla chips, corn curls, and puffs, as well as cheese-flavored and filled crackers make consumers wrongly believe that they are consuming products containing real cheese.

The FDA pointed out that there has been a marked increase in the use of artificial cheese flavors and enhancers in a variety of food products, as well as of enzyme-modified cheese solids to increase flavor intensity. In 1976 the FDA reported that available information "indicates that there are products on the market that appear to be flagrantly misrepresented," since they "fail to contain the valuable constituent that they were represented to contain."

The FDA is at fault for any flagrant misrepresentation of these products. The present state of affairs has been allowed to develop as a result of the FDA's permission to drop the word *imitation* from the label of such products. (See Chapter 15.) The agency's policy created a void regarding terminology for imitation cheeses. Industry prefers the euphemistic term *cheese analogs.*

The National Cheese Institute proposed Standards of Identity to the FDA for products with all or some of the milk ingredients replaced by nondairy ingredients, whereby such "filled" cheese products would be nutritionally equivalent to the cheeses they simulate. A proposed name for such products was *Golana* (*analog* spelled backward). The institute also proposed that certain modifying phrases such as *cheese analog* be used on the label in a smaller type size than the word *Golana,* but the use of any cheese name would be prohibited. The institute requested prompt action from the FDA and FTC "to preclude potential misleading references to cheese in the promotion of Golana."

It is unconscionable for a federal agency to sanction imitation cheeses as "nutritionally equivalent" to traditional cheeses in the absence of proof or in the presence of negative findings. As a result of such official sanctioning, imitation cheeses can now replace real cheeses in the school lunch program, find their way into many retail food products, and be foisted on an unwitting public in restaurants, hospitals, and other places where food is served.

7. Butter versus Margarine

《《《《《《《《《《

Margarine is one of the major successes of food technology. It is a completely artificial product.
—Dr. Magnus Pyke, *Technological Eating, or Where Does the Fish Finger Point?* (London: John Murray, 1972)

Margarine should not . . . be regarded as a perfect substitute for butter. There are no butterfats in margarine, and even if it is fortified with vitamin D_3—not vitamin D_1 nor D_2—it can never be a true substitute for butter. The emphasis must be ever on the natural foodstuffs.
—E. W. H. Cruickshank, M.D., D.Sc., *Food and Nutrition, the Physiological Bases of Human Nutrition* (Edinburgh: E. & S. Livingstone, 1946)

The Paris Council of Hygiene ruled that the artificial product, margarine, was not to be sold under the name of butter. [1873]
—In Reay Tannahill, *Food in History* (New York: Stein and Day, 1973)

The Federal Trade Commission (FTC) has responded affirmatively to a request from the American Butter Institute that the commission look into a possibility that a . . . margarine advertisement is in violation of the FTC Act which prohibits margarine from being represented as a dairy product. According to the ABI newsletter, the ad says "Today's butter . . . is margarine." [1973]
—In *Dairy, Natural and Dietary Food Industry Newsletter*, 12 December 1973

The Great Nutrition Robbery

It's obviously impossible to create foods exactly comparable to natural foods, since we simply don't know everything about nutrients. For this reason, it's unwise to depend too heavily on fabricated and processed foods.

—Dr. Jean Mayer, New York *Daily News*, 20 December 1974

STUDY FINDS MARGARINE MAY TOP BUTTER IN ATHEROGENICITY
—headline, *Medical Tribune*, 8 May 1974

A new margarine substitute: free from chemical additives, based on old family recipe passed down from cow to cow.

—sign in a grocery store, 1975

>>>>>>>>>>

Butter versus Margarine—Consumer Preferences

A survey of nearly 1,400 individuals conducted in 1975 showed that advertisements influenced consumers' choices for margarine over butter. The respondents recalled advertisements and store displays for margarine more frequently than those for butter or any other dairy product.

The survey revealed that more than half the respondents thought that margarine is better nutritionally than butter; the cholesterol issue influenced their choice. They believed that margarine is lower in fat than butter. Obviously, the research findings on hydrogenated fats have not reached the general public: *the fat content of butter and margarine is identical.*

Margarine is apt to be purchased because it costs less than butter. For years, margarine has been advertised as less expensive than the high-priced spread, but gradually the price gap has narrowed. Margarines increased in price by a whopping 63 percent from December 1970 to March 1974, whereas during that same period butter increased only 8.9 percent. In certain markets, some top-priced margarines actually cost *more* than butter.

Numerous questionnaires have shown that butter is still preferred to margarine for taste, despite the skills of the food technologists. In earlier sample testings, reported in 1963, no margarine was equated with top-quality AA 93 score butter. Even the highest-ranking brands of margarine have been judged to have a somewhat artificial flavor, greasier in texture than butter. Most margarines were described as melting poorly in the mouth "with a feeling reminiscent of salve or vaseline."

The History of Margarine Manufacture

Margarine, the earliest nondairy substitute, is a well-known example of a fabricated food. Despite many obstacles, including scientific and technological hurdles, legal battles, and marketing problems, margarine has found public acceptance "not as a poor man's substitute for butter, but as a middle-class alternative."

The search for a butter substitute began in the early 1800s, when butter was frequently rancid and adulterated with lard. Michel Eugène Chevreul, a French chemist, created a product composed of two common fats, stearin and palmitin, and named the product *margarin* (from the Greek *margarītēs,* for its pearly, lustrous appearance). Later another French chemist, Hippolyte Mège-Mouriés, studied butterfat. He believed that the glands of the cow's udder were responsible for converting body fat into butterfat. He imitated mechanically what the cow did naturally. By extracting beef tallow, he obtained a yellow fat and pressed it to obtain a more liquid fraction, oleo oil. He blended it with chopped cow's udder and churned the mixture with milk and water to yield a product that he called *oleomargarine.* Crude and misconceived as this process now appears, it contained the germ of an idea that later developed into margarine manufacture in both Europe and America, with commercial production by 1873.

Regulatory officials anticipated that the new product might be misrepresented. The Paris Council of Hygiene prohibited the use of the term *butter* for oleomargarine, and other countries established similar regulations.

Terminology did not worry American processors. They recognized the potential for a butter substitute and called the new product *but-*

terine. Dairy interests fought back, reportedly by circulating rumors about unsavory origins of the fat used in butterine production. Despite the rumors, sales increased. By 1876, America was annually exporting to the United Kingdom more than a million pounds of butterine.

As acute shortages developed for animal fats used in oleomargarine production, the problem was solved by applying a new process, hydrogenation. Liquid vegetable oils could be substituted for the scarce animal fats. By hydrogenation the oils were converted into hard fats. This invention led to the development of hitherto untouched and seemingly limitless resources of vegetable oils, especially from Africa.

For years, powerful dairy interests succeeded in preventing margarine from being colored at the factory, a pressure that discouraged margarine sales. Homemakers disliked the inconvenience of having to color the margarine at home. The unappealing appearance of the uncolored product resembled cold cream or lard. In addition, the public continued to associate margarine with food shortages during World War I.

Shortly after World War I a bitter struggle began between the butter and margarine interests that continued for three decades. At the beginning, when margarine was manufactured by small processors, it was easy for the butter interests, affiliated with the powerful dairy industry, to dominate and dictate policy. That situation changed as margarine manufacture gradually became affiliated with other food industry giants. By 1952 the sale of colored margarine was legal in all states except Minnesota and Wisconsin, two strong dairy states. Later, even these holdouts had to yield.

The legal sale of colored margarine provided impetus for a phenomenal sales increase of the product. Margarine was sold at a far lower price than butter. Innovative packaging experts wrapped margarine more attractively than butter. Motivational researchers removed the low-social-status stigma and projected margarine as a modern, efficient product, consumed by forward looking, progressive people. The "spreadability" of margarine was emphasized. By 1958 margarine outsold butter by more than 100 million pounds annually. Margarine processing was improved. Manufacturers invested large sums for research in the technique of "creaming" the product, maturing it with microorganisms similar to ones that help create butter, and by using vegetable oils in margarine manufacture.

However, the great bonanza developed when margarine processors began to exploit public worry about cholesterol and coronary disease. The advertising campaign launched by margarine manufacturers was termed "one of the most unprincipled food promotions . . . in the past quarter of a century," with TV commercials described as "noisy, ubiquitous and shameless. They have promoted a staple food as though it were a drug."

Margarine advertisements were directed especially to physicians, who were inundated with promotional literature touting the benefits of polyunsaturated oils used in margarine. The implication was clear: use of these products would reduce cholesterol and prevent coronary diseases. Vegetable oil processors joined margarine manufacturers in this campaign. Physicians, lacking information about how the hydrogenation process affects human health, or about the hazards of too much polyunsaturated fat, began switching patients from butter to margarine and from animal fats to vegetable oils.

The Hydrogenation Process

Hydrogenated fats, which are very durable and do not turn rancid readily, hence are favored by food manufacturers. The oils used to make margarine oxidize on exposure to air and lose their freshness easily. They become rancid much more quickly than saturated fats. To avoid this spoilage, the oils are chemically modified to increase their degree of saturation, by means of hydrogenation. The *cis* forms of fatty acids, naturally present in oils, are converted to *trans* isomers. For the objective of obtaining a durable product with a higher melting point *trans* isomers are helpful. The alteration makes them suitable for the manufacture of margarine and other plastic shortenings, as well as for use in various processed foods including breads, cakes, pies, and cookies. The raised melting point improves the fats' consistency and color in french frying of foods and protects the fats from developing off-flavors or becoming rancid quickly.

However, the original molecular pattern of the vegetable oil has been reorganized and the altered biological quality makes such oil nutritionally undesirable. Hydrogenation changes some of the polyunsaturates to monosaturates. Although the newly formed *trans* isomers may be metabolized as readily as the *cis* fatty acids, the pattern of metabolism is altered. Moreover, dietary fatty acids can be used without metabolism and become incorporated into the structure of the body depot fat and other fat structures.

Trans isomers are not found normally in human tissues. They do not act in the same manner as normal fatty acids in the human body. Hydrogenated fats, with their higher melting point than fats that are liquid at room temperature, are less well utilized in the body. They do not circulate in the blood or move through the tissues as liquids. Hence, they may disrupt the permeability characteristics of the membranes of the body's cells and prevent the normal transport of nutrients into and out of cells.

By the mid-1950s warnings were sounded about hydrogenation of fats and oils and the effect on human health. An article in the prestigious medical journal *The Lancet* predicted, "The hydrogenation plants in our modern food industry may turn out to have contributed to the causation of a major disease."

Dr. Hugh Sinclair, at the Laboratory of Human Nutrition at Oxford University, reported that hydrogenation of fats produced a deficiency of essential fatty acids (EFA) by destroying them, or produced abnormal toxic fatty acids with an anti-EFA effect both in experiments with animals and clinical experiences with human beings. Sinclair's research demonstrated that a deficiency of EFA is "a contributory cause in neurological diseases, heart disease, arteriosclerosis, skin disease, various degenerative conditions such as cataract and arthritis, and cancer."

Sinclair's findings were elucidated by another distinguished Briton, Franklin Bicknell, M.D., who described how hydrogenated fat forms new molecular structures that are unacceptable to the human physiology.

> The abnormal fatty acids produced by "hardening" [hydrogenation] are the real worry. The atoms of the molecule of an essential fatty acid are arranged in space in a particular manner . . . but hardening may

produce a different spatial arrangement, so that a completely abnormal . . . unsaturated fatty acid is produced.

Bicknell compared this different spatial arrangement to ordinary handwriting and mirror handwriting. Although they are identical, they are spatially different. At best, reading the latter is difficult; at worst, serious mistakes may be made. Using this analogy, Bicknell said that the same mistakes are made by the body when it is presented with the abnormal EFA. Not only does the body fail to benefit from them, but it is deluded by their similarity to normal EFA and so attempts to use them. The body tries to incorporate them in biochemical reactions but then discovers that they are the wrong shape. Unfortunately, the reaction has gone too far to jettison them and begin again with normal EFA, so they not only are useless but also actually prevent the use of normal EFA. In fact, they are *anti*-EFA and they accentuate a deficiency of EFA in humans and animals.

Bicknell noted that in World War II in Norway, where margarine factories had been destroyed, arterial diseases decreased. In England, during the same period, with margarine factories intact, arterial diseases increased. Bicknell commented, "It is difficult to resist the conclusion that our increasing arterial degeneration is not the inevitable concomitant of old age, which it may antecede by many years or never join at all, but a *preventable pandemic disease of modern foods and especially of modern bread, milk and margarine*" (emphasis mine).

More recently, the *trans* isomers in margarines were investigated by Dr. Fred Kummerow, professor of food chemistry at the University of Illinois, and his associates. They fed a basic diet to ten groups of pigs with a dozen pigs in each group. All animals received balanced rations with 3 percent fat. This was the only food of the controls. The diets of the nine experimental groups were supplemented by one of the following: beef tallow, rearranged fat free of *trans* fatty acids, corn oil, butterfat, hydrogenated soybean oil with 50 percent *trans* fatty acids, a mixture of used fat and sugar, egg yolks, whole egg powder, or crystalline cholesterol equivalent to two eggs per day.

Of the twelve animals given hydrogenated fat with a high content

69

of *trans* fatty acids, seven, or about 58 percent, had raised lesions of the aorta, which is considered an early symptom of atherogenicity. By comparison, such lesions were found in only 14 percent of the remaining animals.

Animals in this *trans* fatty acid group also showed the highest total cholesterol levels in the blood. The amount of cholesterol found in the aortas did not vary significantly, regardless of the kind of fat in the diet. But the lesions observed in the *trans* fatty acid group had a higher cholesterol content than did the involved aorta areas of animals fed diets supplemented with cholesterol.

If these findings are applicable to humans, it means that *margarine which contains a high level of* trans *fatty acids is more likely to cause atherosclerosis than cholesterol-rich animal fats such as butter, or cholesterol-rich animal foods such as eggs.*

Some margarines produced in the United States contain from 36 to 40 percent *trans* fatty acids, which is slightly lower than the 50 percent level used by Kummerow in his experiments. Some Canadian and European margarines, processed in part from rapeseed oil, may contain even higher percentages of *trans* fatty acids than American margarines.

After Kummerow's findings were reported, the Shortening Institute announced that it was sponsoring a study to check on the possible atherogenicity of hydrogenated fats. The Unilever Company was sufficiently impressed by Kummerow's findings and similar studies of their own to market a *trans*-free margarine in Holland, Germany, Sweden, and Finland.

8. Restructured Fruits and Vegetables

〈〈〈〈〈〈〈〈〈

We are still at the primitive stage of the modern extruder. Manufacturers will perfect machinery and equipment, methods of manufacturing will be improved by the packers, and more attention will be paid to the types of food grown and raised especially for extruding processes.

—*Quick Frozen Foods,* May 1976

We have been very impressed recently, through a number of consumer focus group interviews . . . at the . . . regard for old-fashioned commodity-based products . . . products that are generally regarded as naturally nutritious or "wholesome." . . . This does not mean that people have given up on innovation, but that they do want new things that come from familiar, natural bases, and not things that they regard as technically prefabricated. The rule might be expressed as, "start with the familiar and trusted and then become unique."

—Menaker and Dangerfield Communications Associates, Inc., in *Food Processing,* July 1975

Consumers seem to be turning away from the fabricated products because they perceive them as unnatural, jejune and overpackaged. The nitrogen flushed canister Pringles and the others come in, offers distinct advantages in shipping, warehousing, supermarket display, shelf life and so on. But these are largely designed to benefit someone other than the housewife (the

71

manufacturers, the supermarket management, etc.). To harassed home-makers, now facing more convenience in foods than she can cope with economically, Pringles' "advantages" come high. Currently, largely because of the high cost of the canister, they are selling for about ten percent more than like amounts of real chips.

—*Snack Food,* June 1976

〉〉〉〉〉〉〉〉

Processed fruits and vegetables may now be submitted to harsh treatments of pressure and heat that convert them into convenient forms from which food products are formulated. Such restructured fruits and vegetables bear no resemblance to their original form in appearance, flavor, or nutrient composition.

Pelletized Food

A continuous, completely automated system now converts many liquid or semiliquid foods and food particles called fines into quick-frozen pellets used by food manufacturers to fabricate food products. Initially, the system was used to pelletize chopped spinach, but its use has been extended to a wide range of other items.

Any food that can be made into a slurry or liquid can be pelletized. It is pumped onto two specially shaped belts, one profiled and the other flat, by a spreading mechanism. The flat belt is brought into contact with the product and encloses it totally in a continuously moving "sandwich." The product is frozen and formed. The two belts are then separated. The frozen product adheres to the flat belt, is passed to a final forming operation, enters an "outfood conveyor," and is finished in the form of quick-frozen pellets.

Puff-exploded Food

"Puffed," "texturized," "gelatinized," "homogenized," or "agglomerated" are descriptions of puff-exploded food products. Finely divided food particles undergo high heat, pressure, and steam in a special machine. Food-research engineers in the USDA have devoted considerable effort to producing puff-exploded fruits and vegetables by means of low-pressure puffing guns. Among a number of puff-exploded foods that have been developed are carrots, used by food processors in deli salads, meat seasoning blends, sandwich spreads, garnishes, oriental dishes, instant dry soups, and à la king sauce mixes. Puff-exploded foods are popular because cooking time can be decreased by using them.

Granulated and Flaked Food

Food technologists have developed a technique for converting fruits and vegetables into dry, free-flowing granules or flakes; they view this technique as having unlimited applications. Dehydrated flakes, combined with a modified food starch, may partially or totally replace the pieces of apple in apple pie. The peach filling in Danish pastry or the pumpkin in pumpkin pie or the tomato base in tomato sauce may consist of granules or flakes. Similar applications can be made for berries, carrots, or beets. Granulated or flaked foods lower production costs substantially.

Shredded, Bagged Food

Since produce such as head lettuce deteriorates rapidly, where lettuce is harvested it may now be cored, trimmed, washed, dried, shredded, and packed tightly into plastic bags. The bags are shipped long distances and the produce can remain fresh for up to twelve days. Most of this bagged lettuce is sold to fast food chains. The same treatment is applied to sliced onions, cole slaw, diced peppers, and shredded carrots.

Dehydrated Food

Fresh cabbage is cleaned, cored, shredded, or chopped and then is air dried and compressed. By dehydrating the cabbage the use of packaging material is reduced, the transportation and storage space requirements are lowered, and the cabbage in this form has a long storage life at room temperature. Later, water is added to the dehydrated cabbage shreds or pieces, which plump up and are used for making cole slaw in restaurants. A surfactant is added to the dehydrated cabbage to hasten rehydration.

Extruded Onion Rings

Onion rings, a familiar and popular restaurant item, were made available to the food service market more than twenty-five years ago, in the frozen, raw, breaded state. But this form was impractical for the fast food operations that have developed more recently. Such operations needed an onion ring that could be handled by inexperienced kitchen workers and that, if partially thawed, could still be refrozen without loss of flavor or breading. All these problems were solved by the extruded onion ring.

Onions are washed, peeled mechanically, then diced. They are suspended in a matrix mix that acts as a "skin setting" bath. The blend is pumped into an extruding machine that forms uniform onion ring–shaped items that are then breaded or batter dipped, and deep fried, blast frozen, packaged, then shipped to restaurants.

When first introduced, extruded onion rings were viewed by many food service operators quite cautiously. "This really isn't an onion ring. It doesn't have a whole onion in it. It doesn't have the consistency or bite our former ring had." But fast food chains quickly recognized the benefits of using extruded onion rings. By adding them to menus, overall sales and profits were increased. Major fast food chains report that this one food item accounts for 3 to 4 percent of their total sales volume, and with special promotion it can go up to 8 percent.

Some extruded onion rings are made from dehydrated onions, first reconstituted in water, then added to the matrix mix. The use of dehydrated onions lowers production costs.

Consumer acceptance of extruded onion rings has led to their retail sale. At present, about half of all onion rings sold in stores are extruded products. Consumers are not always able to distinguish between traditional and restructured onion rings. To clarify the differences, the FDA formulated regulations. Since 1 December 1977, fabricated onion rings must have label statements "made from diced onions," "made from fried diced onions," or "made from dried diced onions."

Restructured French Fries

Processors take whole raw potatoes apart, cell by cell, and blend them with other ingredients, including ascorbic acid. The mixture, reshaped into uniform crinkle-cut french fries, is frozen and shipped to feeding institutions, especially fast food places and schools. The product is described as "uniform in flavor, texture, color and appearance twelve months a year because all the potato variations have been controlled. Using this process it is possible to control the preservation of vitamin C to meet the requirements of the school lunch program."

Restructured Potato Chips

More than a century ago, George Crum, an American Indian who worked in a hotel kitchen at Saratoga Springs, New York, created a food item by deep-frying raw potatoes that had been sliced paper thin. The novelty product, called a potato chip, continued to be made this way. The dictionary definition came to be accepted by the FDA.

By the late 1960s, however, with the creation and marketing of a reconstituted potato chip from dehydrated potatoes, the understanding of what constituted a potato chip became unclear. A bitter struggle followed between producers of traditional and reconstituted chips, with the former claiming that the latter's product changed the basic integrity of chips.

Traditional chips, random shaped, break easily and cannot be packaged compactly. Such products have a limited shelf life, becoming rancid in a month or two. For these reasons, traditional potato chips were produced regionally and distributed by relatively small companies at frequent intervals to insure freshness of the chips. Reconstituted chips were devised to be durable. Being uniform in size and shape, they could be stacked efficiently in cylindrical containers to be shipped anywhere and, combined with preservatives, have a shelf life of up to a year.

The introduction of the restructured chip came at a time when traditional chippers had problems. With high food prices in the early 1970s, consumption was declining for many snack foods, including chips. At the same time, the cost of raw ingredients increased greatly. Despite the decline in sales of snack foods, the restructured chip, introduced by promotional steamroller techniques, sold well as a novelty item. And the dehydrated potatoes used for the restructured chip could be of cheaper grade than the fresh, raw potatoes used for the traditional chip.

Traditional chip makers accused their opponents of unfair competition. Consumers had been told that traditional chips "don't stack up," which could have been taken literally, or have meant that traditional chips were not fresh. By the chippers' own admission, some traditional potato chips were on the shelves too long. Traditional chippers engaged in counteradvertising, ridiculing the ersatz product and noting the extensive use of chemical additives and preservatives in these products. Traditional chippers also sought legal action to stop the use of the term *potato chip* for restructured chips. They contended that the time-honored term should be reserved for their products and others could be designated by another term, such as *potato snack*. The case and an appeal were both lost. The trade organization Potato Chip Institute International (PCII) then turned to a costly campaign to influence Congress and the FDA. The group gathered signatures of consumers and potato growers and conducted a consumer impression survey requested by the FDA to help decide, "What is a potato chip?" The survey showed that 85 percent of the consumers of restructured chips thought that they had actually purchased traditional ones.

The FDA delayed any decision for three years. When it formulated its definition, it tilted heavily in favor of the processors of reconstituted chips. It contended that, by definition, the potato chip is thin, fried crisp, and at times salted. Within that context, the FDA reasoned that *both* the traditional and restructured products "have many more characteristics in common than they have differences" and that "the unqualified use of the term *potato chip* is not false, but rather incomplete . . . additional words are necessary to describe the food informatively." The FDA proposed to consider restructured chips as potato chips with the proviso that the phrase "made from dried potatoes" appear in type at least half the size of the words "potato chips," and the regulations would become effective only two years later, at the end of 1977.

The traditional chip manufacturers resented the FDA's proposal, which they judged "ill conceived, arbitrary and unlawful." The ruling came nearly three years after the original petitions had been filed; meanwhile, the agency had permitted "mislabeling" and "misbranding" to continue. The editor of *Snack Food* commented: "It is to the advantage of the manufacturers of these simulated products to have the decision delayed as long as possible. This gives them additional time in which to use the term potato chip as long as possible and to establish their products in the public eye. It is already past the point, we believe, when sales would be dampened by a ruling which would require them to print the term 'made from dehydrated potatoes' or some such terminology, in larger type . . . than is presently the case." Then, with a tone of philosophical resignation, the editor added, "However, better late than never."

Conventional chip makers viewed the FDA's decision as "justice delayed, justice denied" and appealed, but the FDA refused to rescind the label order.

How Do Restructured Chips Compare with Traditional Chips?

Although both types of chips are high-priced snacks compared to the basic raw commodity the potato, restructured chips are more costly than traditional ones. In 1974 restructured chips sold for 9.6 cents per ounce, compared to 8.9 cents per ounce for traditional ones.

Snack Food conducted a survey, asking top grocery chain executives and buyers about their observations. Among reactions, it was noted that "kids love to look for chips of different color and size," which they can't do with the uniform extruded chips.

Regarding flavor, among some sixty-eight respondents to the survey, there was an even standoff. But the negative comments were strong: "lousy product," "product does not have true potato flavor or quality," "flavor just isn't there," "the flavor is obviously artificial," and "they do not have a natural potato flavor."

According to *Snack Food*, "Many consumers perceive restructured chips as make-believe products in an era forever gone, developed during the age of innocence in the 1960s, when the blandness, uniformity and 'grooviness' (stackability) of such products were held in awe. Today, such products face a more distrustful public. . . ."

9. Real versus Imitation Fruits and Vegetables

《《《《《《《《《《

Factory-made foods are technological achievements that appeal to the tech-nostructure.

—John Kenneth Galbraith and M. S. Randhawa, *The New Industrial State*
(Boston: Houghton Mifflin, 1967)

Dr. J. P. Stockton, a food management consultant, . . . suggested that con-sumers get a safer product when it comes from a factory. "We know every-thing that goes into a created product," he said. "And everything is tested. It's a lot different than a natural-grown product that falls out of a tree into an area where cows are roaming and birds are nesting. These nuts or fruits are picked up by hand and processed for human use. The factory environment is a lot cleaner."

—*Virginian-Pilot*, 8 June 1977

The "stabilized blueberries" in the blueberry pancake mix were purple pel-lets consisting chiefly of sugar, gum acacia, citric acid, starch, artificial color, artificial flavor, and blueberry pulp. The charge was "false and mis-leading labeling."

—*Notices of Judgment* under the Federal Food, Drug and Cosmetic Act,
August 1960

Ingredients in a "blueberry bud" contained in a blueberry waffle mix: sugar, blueberry solids with other natural flavors, vegetable stearine (a release agent), salt, blackberry solids, methyl cellulose (a thickening agent), silicon dioxide (a flow agent), citric acid, modified soy protein.

—*Consumer Reports,* October 1975

Tomato replacer: There is no acid taste or color loss, and just two ounces of the concentrate approximates the flavor and aroma of 30 bushels of fresh tomatoes. . . . The new product contains potato and wheat flours, dextrose, salt, sugar, artificial and natural flavors, propylene glycol alginate, certified food colors, and up to two percent of silicon dioxide as an anti-caking agent. The product can then be used as a one-for-one direct replacement for natural tomato.

—advertisement cited by John Keats in *What Ever Happened to Mom's Apple Pie?* (Boston: Houghton Mifflin, 1976)

General Mills' Baron Von Redberry Cereal: sample KF-5922: The name is misleading for a product that contains artificial flavoring and artificial coloring but no berries.

—*78th Report on Food from Connecticut Markets and Farms* (New Haven: Connecticut Agricultural Experiment Station, 1973)

Taste like orange juice, texture like orange juice, aroma like orange juice, just as much vitamin C.

—advertisement for an imitation frozen orange juice, *Saturday Evening Post,* June 1963

How do you get the nonscientist to grasp that synthetic products often make the best sense because they can contribute something to the quality of life that can't be obtained from the natural product? . . . The fact is . . . while some people sneer at Tang, it delivers a lot more vitamin C reliably than fresh orange juice—especially juice that's been shipped from Florida in canned or refrigerated single-strength form.

—A. S. Clausi, Research, General Foods, quoted by Milton Moskowitz in "Nutrition's Nice, But Will It Sell?," *San Francisco Chronicle,* 21 May 1972

The public has very low taste perception. . . . Insufficiently low, it would appear, to prevent the market from foisting upon us an "orange" drink which

Real versus Imitation Fruits and Vegetables

is 100 percent synthetic, none of whose ingredients has ever been nearer to an orange grove than a chemical factory.
—Waverley Root, "Taste Is Falling! Taste Is Falling!,"
The New York Times Magazine, 16 February 1975

The awful fact is that some of the fabricated products taste better than their real-life counterparts. I drank Orange Plus for a year before discovering that it was labeled imitation juice.
—Jean Carper, "Fake Foods: Low in Nutrients, High in Price,"
The Nashville [Tennessee] *Banner,* 30 January 1975

〉〉〉〉〉〉〉〉

The idea of synthetic fruit may be a shocker, but such fabrications have already been created. In 1946, a synthetic cherry was devised and patented and is now marketed successfully in many countries, including Holland, France, Italy, Switzerland, Finland, and Australia. A similar machine-made cherry produced in the United States is offered conveniently in five sizes, from "buckshot" up to three-fourths of an inch in diameter; it is used mainly by bakers.

The fabricated cherry is formed by allowing drops of a mixture of suitably colored and flavored sodium alginate solution to fall into a bath containing an appropriate calcium salt (usually calcium chloride). A "skin" of insoluble calcium alginate forms on the outside of each drop. The drops are allowed to cure, the calcium ions gradually penetrate into the center, and in time the structure gels.

Synthetic cherries are low in cost, uniform in appearance, always available, stable, and durable. They are unaffected by the heat of baking and are used in fruit cakes and pie fillings.

Another example of an achievement in food technology is provided by a patent issued to the General Foods Corporation in the late 1960s for the production of synthetic fruits and vegetables. The product was described as an "edible, crisp, chewable non-uniform agglomerate of calcium alginate cells with the addition of artificial flavorings and other substances." The process was similar to the one for synthetic cherries but broader in scope, to simulate many fruits

81

as well as vegetables. The purpose of the invention was described as a method of preparing artificial fruits and vegetables that may be heated and cooked without losing their characteristic crisp texture. Dr. Magnus Pyke pointed out, however, another aspect of these products: "their almost total lack of food value."

Raisin-flavored Granules

It is impossible to know whether you are picking out a real raisin or only a raisin-flavored granule from a cookie or muffin these days. The raisin crop has declined and the cost of raisins has risen. Raisin-flavored granules, much cheaper than raisins, have been developed to replace or extend raisins in a variety of food products, including cakes, sweet rolls, cookies, breads, cupcakes, and muffins. The granules consist of sugar, corn syrup, vegetable oil and fats, and other ingredients, including raisin flavors that can be either natural or artificial.

Imitation Tomato Solids and Pastes

In the last few years, several factors combined to create a strong demand for tomato solids. Soaring meat prices led to an increased popularity of tomato-based pasta as well as tomato-based rice dishes. Tomato-based heat-in-skillet casserole mixes, introduced and successfully marketed, also increased demands for tomato solids. At the same time, domestic tomato powders, commonly used by processors in casserole mixes, became virtually unavailable, whereas imported powders became very expensive. The time was propitious for developing imitation tomato solids and pastes and stretching them still further with imitation tomato flavors and enhancers (which intensify flavors), and synthesized tomato flavors and tomato flavor oils.

Imitation tomato extenders, generally made from a blend of starches, acids, colors, and tomato flavorings, are used with soups, soup mixes, pizza, sauces, dressings, tomato pastes and purees, catsup, juices, dips, spices, snack foods, bloody mary and other beverages, extruded products, and seasoning blends. Such tomato extenders may replace as much as half the tomato solids in products. Obviously, such extenders lack the nutrients in real tomatoes, even though they replace the bulk and viscosity associated with tomato solids. Also, tomato extenders, lacking in flavor, require the addition of flavoring agents—in the technologists' terms, to "build up," "round out," or "stabilize" the "flavor profile of changing tomato products." Only two ounces of synthesized tomato flavor are necessary to approximate the flavor and aroma of thirty bushels of ripe tomatoes.

Real versus Synthetic Orange Juice

Since the early 1960s a full-scale war has raged over the question, "When is orange juice not orange juice?" At first, the main adversaries were trade associations from the Florida and California citrus industry. California oranges are more acidic than Florida ones. A California company created an orange juice blend that consisted of California and Florida oranges together with sugar, water, fruit solids, and other ingredients. The product contained only 70 percent orange juice, and the Florida citrus industry objected. The FDA, responsible for creating and maintaining food standards, was caught in the crossfire.

Consumers erroneously believed that they were buying 100 percent pure orange juice with products labeled "orange blend," "orangeade," "orange nectar," "orange juice cocktail," "orange juice drink," or "orange drink." To add to the confusion, some of these products were packaged in a way similar to 100 percent orange juice, and some processors began to can frozen concentrated imitation orange juice as well as pure frozen orange juice concentrate, in similar cans.

Even the President's consumer advisor, Virginia Knauer, admitted that she was confused. She reported her experience to the National Juice Products Association, a trade group representing producers of imitation orange juice products. "I had no idea of what I was buying, no criteria for understanding the differences in quality.

. . . Imagine the plight of the average consumer, when a Presidential consumer adviser is bewildered!" she confessed. Mrs. Knauer told the group that consumers, without standards of identification for these products, will "wonder what they've been paying for all these years, flavored water or orange juice."

In 1964, producers, consumers, and FDA representatives began to discuss Standards of Identity for these products. These discussions turned into wrangling and dragged on into the 1970s. Meanwhile, the problem became more complex. It was no longer merely a question of pure versus diluted orange juice, but also real versus synthetic. Citrus juices faced stiff competition from a proliferating array of substitute and synthetic drinks available as powders or frozen concentrates, containing no natural citrus solids.

Data on the consumption of orange drinks and synthetic orange drinks from 1960 to 1962 are unavailable. During that period, however, frozen concentrated real orange juice held approximately 76 percent of the citrus beverage market. By 1969 its share had declined to 53.4 percent. Since the market for all types of juice increased, it is impossible to interpret from these data the inroads of the substitutes and synthetics. Imitation orange drinks, which are less expensive, undercut the market for real juice. Some consumers thought that the less expensive synthetic orange drinks tasted as good as real fruit juices. Food technologists claimed that children raised solely on synthetic orange drinks actually found the taste of real fruit juice strange or "funny."

General Foods Corporation versus the Florida Citrus Commission

Advertisements for synthetic orange drinks highlighted the fact that the products contained large amounts of Vitamin C (ascorbic acid), sometimes even more than that found in real orange juice. Television advertisements were denounced by citrus industry spokesmen as "attempts to instill in the mind of the viewer that pure orange juice doesn't contain any taste-pleasing factors," and the commercials were described as "a blatant attempt to derogate the natural orange juice product for the financial gain of an orange additive product." One commercial portrayed a child saying, "I hate orange juice." The citrus industry called these advertising attacks a program of vilification.

The citrus industry fought back. The Florida Citrus Commission paid three nutritionists $56,000 to prepare a report which the commission hoped would point out the nutritive values of citrus juices and which the commission could use in counteradvertising. The recommendations of the committee were disappointing and played directly into the hands of the processors of synthetic products. The report downgraded the importance of natural Vitamin C. All three members expressed the view that this vitamin, created in the test tube for synthetic foods, is as effective as the natural vitamin found in real foods. (For refutation of this claim, see Chapter 13, Natural versus Synthetic Vitamins.)

The Florida Citrus Commission launched its own advertising campaign. One advertisement ridiculed those "imitation juice products which have a list of the funniest sounding chemicals" on their labels, and another announced that "Mother Nature provides her children with genuine Florida orange juice. Don't let chemical imitations persuade you to do less for yours."

A crucial battle was fought by the two warring camps in 1969. General Foods Corporation advertised several of its orange-flavored drinks as pure orange products. For several months, Florida officials fought back unsuccessfully. Then Governor Claude Kirk, Jr., of Florida ordered all Florida state agencies to discontinue purchasing any products manufactured or distributed by the General Foods Corporation. The Florida Citrus Commission, assisted by the state's congressional delegation, sought legal redress from the Federal Trade Commission, but the FTC announced that the practices under investigation of alleged "false and misleading advertising in connection with the sale of orange-flavored drinks have been discontinued and will not be resumed."

General Foods had signed a voluntary compliance document that had led the FTC to drop further action. The document stated that the corporation "agreed to no longer represent directly or by implication that Orange Plus or other fruit-flavored concentrates or beverages are products of natural orange juice or would make natural juice when reconstituted." General Foods agreed to disclose clearly and conspicuously on its label and in each advertisement that, in fact, Orange Plus is a frozen concentrate for imitation orange juice and is not a wholly natural juice product. In addition, "advertising claims will not use disparaging or derogatory statements concerning the nature or quality of any natural orange juice." (Although General Foods' products Tang, Start, and the frozen concentrate Awake contain no orange juice, the product Orange Plus contains 50 percent orange juice and added water. Today the label reads "frozen concentrate for imitation orange juice.")

Standards of Identity for Watered-down Fruit Juice

Meanwhile, those discussions of Standards of Identity for orange juice blends, begun in 1964, dragged on for seven years. Confusion continued to reign in the marketplace and was described by Dr. Jean Mayer as "an outright scandal."

The citrus industry was willing to label all orange drinks with their specific percentage of orange juice. This no-nonsense offer would have clarified label information for consumers, but the FDA rejected the suggestion, arguing that it would be too difficult to police. Instead, it proposed to establish four broad categories of orange juice content: (1) orange juice drink blend, 70 to 95 percent orange juice; (2) orange juice drink, 35 to 70 percent orange juice; (3) orange drink, 10 to 35 percent orange juice; and (4) orange-flavored drink, not to exceed 8 percent orange juice. Increments in the first three categories were to be shown in units of five. For example, if a product was in the first category of orange juice drink blend and contained only 71 percent orange juice, the label could read 75 percent. For the last category, orange-flavored drink, the increments were to be shown in units of two.

The citrus industry's proposal would have been more informative and less complicated for consumers than the FDA proposal. Some consumer groups criticized the FDA's categories. Under this plan, if one brand has only 35 percent orange juice and a competing brand has 65 percent, both fall within the 35 to 70 percent category, though the values of the two products differ considerably. This arrangement offers producers no incentive to add more real fruit juice and, in fact, encourages them to congregate at the very bottom of each category.

The FDA argued that laboratory tests cannot determine a product's exact juice content, even within 10 percent. The FDA expressed concern that the industry proposal would spark "juice power" labeling competition, with the agency unable to know whether the producers' claims were true.

Nutritional Equivalency

How do the orange juice blends and synthetic orange products compare nutritionally to real orange juice? Claims for the synthetics stress their Vitamin C content, as discussed earlier, but there are distinct differences between synthetic ascorbic acid and the Vitamin C complex found in foods. Real orange juice contains fractions of the Vitamin B complex: niacin and thiamine, which may be totally absent in synthetics. Orange juice, an excellent source of potassium, contains more than four times as much potassium as the synthetic fruit drinks. Doubtless there are many other important nutrients, including other vitamins, minerals, trace minerals, bioflavonoids, and fiber, that are well balanced and available in real orange juice but not yet listed in nutrition charts.

The value of the blended drinks rises, naturally, with the higher percentages of real fruit juice they contain. But they also contain undesirable ingredients such as added sugar and artificial flavors and colors.

Liquid and dry bases, syrups, concentrates, and premixes, made with no real fruit and appealing largely to young children, are especially objectionable. Parents need to be made aware of the true nature of these products, which include a host of artificially flavored drinks: orange, lemon, grape, and punch. One product, typical of many, is a pink lemon-flavored base described nostalgically in its advertisement: "Remember how great Pink Lemon used to taste? It still does . . . hard to miss that . . . 'old-time' flavor . . ." The ingredients of that pink lemon-flavored base are cane sugar, citric acid, dextrose, natural flavor, oil of citrus fruits, sodium citrate, calcium phosphate, gum arabic, Vitamin C, U.S. certified color, and butylated hydroxyanisole. Old-time flavor?

10. The Staff of Life versus Carbohydrate Substitutes

《《《《《《《《《

What emerges from the corporate baker's oven is not precisely bread. It is something that . . . has [been] . . . described as "a ghastly cloud of sugared chemistry."

—John Keats, *What Ever Happened to Mom's Apple Pie?*
(Boston: Houghton Mifflin, 1976)

Potatoes are . . . largely, and perhaps constantly, used by fraudulent bakers, as a cheap ingredient, to enhance their profit. . . . This adulteration does not materially injure the bread. The bakers assert that the bad quality of the flour renders the addition of potatoes advantageous as well to the baker as to the purchaser, and that without this admixture in the manufacture of bread, it would be impossible to carry on the trade of a baker. But the grievance is, that the same price is taken for a potato loaf, as for a loaf of genuine bread, even though it must cost the baker less.

—Frederick Accum, *Death in the Pot* (London: circa 1830)

According to Harvey O'Connor's book *Mellon's Millions,* published in 1933, Mellon fellows were supported by grants from individual companies to do confidential research [for] their sponsors. . . . In [one] project, a secret process for bread-making which reduced by one-half the needed amount of yeast and sugar, result[ed] in increased profits for the company of a million dollars a year. . . . Nutrition experts shook their heads. White bread, they declared, depends on yeast for its vitamin content. If baker's bread is tasteless and lacking in nutrition, it is due partly to science's contribution in making available the use of cheaper grades of flour and savings in other constituents in the staff of life. . . .

—Rachel Scott, *Muscle and Blood* (New York: E. P. Dutton, 1974)

A versatile, transparent plastic, biodegradable and independent of petroleum feedstocks . . . [is] starch-derived polymer pollulan. . . . [The processor] is looking closely at food packaging applications—and at pollulan as a diet food itself. The material . . . is inert in the body and is excreted completely. . . . As a diet food, pollulan can be used instead of flour in baking bread and pastries. . . . It has virtually no calories at all.

—*Chemical and Engineering News*, 24 December 1975

A new bread, made from cellulose fibers will soon be on the market. . . . The new bread contains 25 percent less calories than most breads—yet has 25 times more roughage. . . . [The announcement] failed to mention how the bread tastes.

—*Food Management*, March 1976

Cellulose provides no nourishment to the eater. Once it gets inside the body it takes up the same space as food, thus helping to satisfy hunger pangs, but since it is not absorbed as food it does not add any weight. . . . It is not tasty, it is not nutritious, it is totally lacking in calories and vitamins, and in fact it has no food value whatsoever. What it does do is keep the stomach from feeling lonely.

—*Life*, 2 June 1961

Can a bread . . . made with purified wood pulp promising weight loss and better bowel movements find success in the consumer world? . . . According to . . . the developers of the bread, replacing some of the flour and shortening found in traditional bread formulas with powdered wood pulp or cellulose, a purified plant fiber, is accepted for use in food products by the FDA.

—*The Express* [Easton, Pennsylvania], 8 September 1976

The Staff of Life versus Carbohydrate Substitutes

A compound that smells like fresh white bread has been isolated . . . at the Central Institute for Nutrition and Food Research, in Zeist, the Netherlands. The compound is 2[(methyldithio) methyl]furan. The Dutch scientists isolated it from an extract prepared from the crusts of 200 loaves. There ought to be plenty of uses for a compound that smells like fresh bread. This one, however, "decomposes easily into dimethyl disulfide and difurfuryl disulfide," which seem unlikely to be treats for the nose.

—Chemical and Engineering News, 13 September 1976

A [1958] Japanese patent . . . approves a rice substitute made of a mixture of starch and flour (2:8). This is heated with water to 120°C. for ten minutes, dried and shaped into grains like rice. The grains are then treated with dibenzoylthiamine hydrochloride; carboxymethyl cellulose is added and they are ready after five hours.

—Modern Nutrition, March 1963

〉〉〉〉〉〉〉〉

An interest in synthesizing carbohydrate—the main component in bread—has existed for more than a century. Recently, with the rising costs of carbohydrates, there is increasing interest in developing alternative bulking agents for foods and replacing traditional flour and sugar. The starch in flour closely resembles cellulose, the structural component of wood. However, whereas the starch in flour is edible by humans and other nonruminants, cellulose is not.

Cellulose is derived from readily available plant materials, especially wood and paper waste, and yields dextrose through enzymatic conversion. In turn, it can be converted into secondary products such as high-fructose corn syrup, or alcohol, by fermentation. Numerous cellulose extenders, bulking agents, and other functional substitutes for carbohydrates have been developed.

Microcrystalline Cellulose as a Bulking Agent in Food

The use of cellulose as a nonnutritive food component was explored in the early 1900s by Frederick Hoelzel, an imaginative dieting enthusiast who searched for bulking agents. In his early trials he ingested such exotica as glass beads, feathers, powdered coal, steel balls, rubber, and short lengths of silver wire. Hoelzel obtained the best results with moist sea sand seasoned with salt. Later he tried cotton batting, which is nearly pure cellulose. Within a year he developed the art of swallowing the batting and discovered that "two ounces of chopped-up surgical cotton flavored with fruit juice" usually satiated hunger pangs for several days. By 1919 Hoelzel produced cellulose flour, but people were not yet ready to accept cellulose as delectable fare.

A series of similar experiments by other researchers proved fruitless. Then the prospects were brightened, by sheer chance, by a chemist, Dr. O. A. Battista, who was attempting to perfect a stronger tire cord in a rayon and cellophane factory. Battista believed that if he could break down bundles of cellulose molecules into tiny fragments, the tire cord would have strength comparable to that of steel wire, but he repeatedly failed to obtain the desired tiny cluster of molecules. Then he mixed a thick batch of cellulose and water in an electric blender, thinking that the chopped-up cellulose bits would fall to the bottom, separated from the water. Instead, the blend turned into a thick gel. Battista saw its potential as a non-nutritive bulking agent with many food applications. He mixed a batch of cellulose cookies, baked them in the laboratory oven, and sampled them. In this manner the non-nutritive bulking agent for food, cellulose, was born.

Microcrystalline cellulose, a highly purified form, exists as minute separate particles. The snow white, free-flowing powder is odorless, tasteless, and noncaloric. It contributes gel stability, body, bulk, opacity, and texture to many convenience foods and makes possible the formulation of many new and highly profitable low-calorie foods. For example, microcrystalline cellulose substituted in part for flour or instant potato flakes reduces the energy value of baked goods and mashed potatoes intended for low-calorie diets. It also offers substantial savings to food and beverage processors by replacing more costly ingredients. In salad toppings, puddings, soups, and desserts, the use of microcrystalline cellulose sharply reduces the needed starch and oil. It can also replace a sizable portion of ingredients in candies, pretzels, and other snack items without adversely affecting flavor.

Wood Pulp in Bread

In the 1970s the dietary-fiber craze struck. Bakers responded to the public interest by formulating specialty breads with added bran as the source of dietary fiber. Soon, however, powdered cellulose—an inexpensive and readily available by-product of the wood pulp industry—was used as the source of fiber in some of these high-fiber breads. Since powdered cellulose has the capacity to absorb and hold a considerable amount of moisture, the main ingredient in these breads is water. Although some flour is present as an ingredient, a considerable portion of it is replaced by less expensive, non-nutritive powdered cellulose, yet the specialty loaf commands a higher price than traditional bread.

Breads containing powdered cellulose are touted as having 400 percent the fiber of whole wheat bread. Is this desirable—or safe? At present, we don't know how much dietary fiber is required by humans; some may be beneficial, more may be harmful. Although naturally existing fiber from grains, fruits, and vegetables may be desirable in the human diet, the important nutrients associated with fiber in such foods are not present in an isolated product such as powdered cellulose.

Excessive fiber intake can result in health problems because it can bind fats, cholesterol, and minerals and interfere with the absorption of essential micronutrients, including zinc, copper, and magnesium. Eating excessive fiber over a period of time can result in the loss of important nutrients and lead to nutritional deficiencies. Animal studies have shown that excessive fiber in the diet induces caloric inefficiency.

The FDA approved the food use of powdered cellulose, despite the fact that it is not known if the human body handles the substance in the same way in which it handles dietary fiber from traditional food sources. Marian Burros of the Washington *Post* tried to find out what tests had been conducted to assure safety before breads containing wood pulp were marketed. She found that two studies had been conducted by Dr. Olaf Mickelson, professor of nutrition at Michigan State University, one for fifty-three days with rats and another for eight weeks with men and women. A third study, at the Mayo Clinic, lasted three months. All three studies were short range.

Burros asked Mickelson if he could give assurances that eating the high-fiber bread for a period of two years would not create micronutrient deficiencies. Mickelson hedged: "As a scientist, I will say this depends on two things—one, the amount of the bread the individual consumes regularly every day, and two, the composition of the rest of the diet." Mickelson admitted that his rat studies were for a relatively short time and that the studies with human beings involved adults, whereas the situation would be quite different for growing children.

Dr. Donald Oberleas, chairman of the department of nutrition and food science at the University of Kentucky, told Burros, "Nobody is really sure if it is dangerous. Nobody is really sure if wood pulp fiber is different from fiber found in other foodstuffs. Or how effective it is. Fiber is such a relatively new area." Oberleas was concerned about the lack of long-term safety data: "Twenty years from now we'll look back and find something has gone wrong. Overindulgence is always a bad policy."

Two FDA representatives interviewed by Burros also expressed doubts, in spite of the fact that the agency granted GRAS (Generally Recognized as Safe) status to powdered cellulose. Dr. Allan Forbes, acting associate director of the Office of Nutrition and Consumer Sciences, admitted being "personally uneasy about what kind of studies were done. . . . The problem comes, when large amounts of fiber are consumed" because they "may affect the absorption of trace minerals." Forbes admitted that no one knows how much fiber is too much: "I doubt there would be any problem [in eating high-fiber bread] but I'm not sure."

Dr. Barbara Harland, a research biologist and fiber expert in FDA's division of nutrition, commented on the inadequacies of the short-range studies conducted. Harland noted that zinc deficiencies do not show up for at least six months and then appear when subjects are placed on diets designed specifically to lower the zinc levels. Dr. Harland added, "Wood pulp from a tree has been okay for animal feeding, but we don't know if it's okay for humans." She recommended getting increased fiber by eating more whole grain breads and cereals, fresh fruits, and vegetables.

The food use of wood pulp as powdered cellulose is not being limited to high-fiber breads. Consumers now stand on the threshold of having this inexpensive waste product incorporated into a wide variety of processed foods and beverages as a substitute for more costly nutritious ingredients, and they will have no idea that they are ingesting wood pulp.

Processors are being wooed. One large paper mill sells powdered cellulose for food and beverage use, describing its product as a "purified plant fiber" made "from nature's own ingredient." Processors are told that the powdered cellulose is a unique bulking agent for flour-based products such as breads, cookies, cakes, breakfast cereals, and pastas. Used in bread coatings for chicken, fish sticks, and onion rings, powdered cellulose increases their crispness. As a binder in snack foods such as chips it prevents disintegration during processing, strengthens thin chips intended for dips, and provides chewiness. With processed meat products such as hot dogs, sausages, ground meat patties, and deboned meat it helps retain fat in the products. In sausage casings it adds strength as a binder and filling. In prepared liquid foods such as salad dressings, sauces, and soups, it absorbs excess water and gives a smooth texture to liquids by dispersing small particles throughout. Additional suggested uses for powdered cellulose by the paper mill include many other food and beverage applications. The product can be incorporated readily into stuffings for meat, fish, and fowl; mashed vegetables; cottage cheese; yogurt; fish cakes; and soft drinks. In the words of the paper mill's producers, the food and beverage uses of powdered cellulose "are limited only by one's imagination."

Various approaches are being used to exploit cheap and readily available substances as flour replacers in baked goods. Many of these have no nutritive value and the safety of their long-run consumption by humans is unknown.

Xanthan Gum: A Flour Replacer in Baked Goods

Xanthan gum, derived from bacterial action on glucose, can replace gluten in a variety of products, can be applied in the manufacture of doughs and batters for dietetic and fast foods, and can be used in prepared food mixes, snacks, and other products made from starch and soybean protein. Xanthan gum is used in high-protein breads, foods for people allergic to gluten (those suffering from celiac disease or schizophrenia), and baked goods in areas of the world where bread wheats are not available.

Doughs made with xanthan gum don't require kneading; a four-minute mixing period is usually enough. Xanthan gum improves the cohesion of starch granules and produces a breadlike structure comparable in appearance, mouthfeel, and staling to most commercial breads. Experimentally, breads baked from cereal starches, without xanthan gum, failed to rise. They remained flat, brittle, and coarse in texture. However, the addition of xanthan gum brought about remarkably good changes in loaf volume and crumb structure.

Additional studies on the use of xanthan gum in other bakery products are in progress. Preliminary studies show good results from use of the gum in cakes, buns, pancakes, and doughnuts.

Pollulan: A Flour Replacer for Baked Goods?

Pollulan, a starch-derived plastic, has been developed in Japan by a subsidiary of the country's largest processors of starch syrups. The company searched for a microorganism that would produce a suitable polymer from starch, and then tried to develop a purification technique for inexpensive mass production. The microbe search led to a yeast strain, *Pollularia pollulans*, and a successful procedure to separate the spent yeast from the product, named *pollulan*. The product is a versatile, transparent plastic that can be substituted for flour in baking breads and pastries and serves as a food film for packaging.

Powdered pollulan is inert in the body and is excreted completely. According to the product development director, pollulan as a diet food can be used as a flour replacer. "It tastes the same as bread, but it's not simply a low-calorie food—it has virtually no calories at all."

Chicken Feathers: A Flour Replacer?

Experimentally, cookies have been made using 10 percent chicken feathers as a flour replacer. At the University of Georgia, "chicken feather cookies" were taste tested by a panel of fifteen. Only one person thought that he could distinguish them from cookies made from traditional ingredients.

The chicken feather distillate is mainly protein. Dr. A. L. Shewfelt, head of the university's food science department, suggested that if masking agents such as raisins and nuts were included in the cookie recipe, the feather powder could replace even more than 10 percent of the flour.

Like wheat flour, feather powder is not a complete protein. It is short on lysine, one of the eight amino acids the body cannot manufacture.

Other Replacers in Baked Goods

New types of dough conditioners and softeners make it possible for bakers to use vegetable oils instead of shortenings and fats. The savings resulting from these substitutions are significant. Liquid oils have always been less costly than plastic fats, and only 2 percent oil will do the work of 3 percent plastic fat.

Milk simulators are being used to replace nonfat dry milk, formerly used in baked goods. The replacers are reported to offer substantial savings of up to 50 percent. The substitutes are made of partially demineralized and partially delactosed whey protein concentrate and a modified food starch. In bakery applications, the quantity of the blend required ranges from 30 to 50 percent of the original quantity of nonfat dry milk. The reduced quantity is adjusted by adding flour or flour and sugar. The replacer, used successfully in commercial bakery operations, can totally replace nonfat dry milk in high- and low-fat cakes and in raised doughnuts. The replacer is also being used in muffins, biscuits, and pancake mixes, snack cakes and cookies, and cream fillings and icings.

The growing trend of using more and more food ingredient replacers not only for bakery goods but with all food products deserves serious consideration. Food technologists appear to be driven by some blind desire to develop replacers. Naturally, processors find replacers attractive when they help reduce production costs. What seems to be ignored is a question that should be primary: how nourishing are substitutes compared to the traditional components they replace?

11. Creations by the Food Flavorists

Flavor makers quietly are conceding that artificial flavors are the wave of the future.
>   —*Wall Street Journal,* 21 November 1977

Looking for new artificial flavors? Talk to one of our great imposters.
>   —advertisement, *Food Technology,* September 1975

Bananas at the ripe stage owe their taste values to over 150 substances—which means that synthetic banana flavors are hardly a match for the "real thing."
>   —R. Tressl and F. Drawert, *Journal of Agricultural and Food Chemistry,*
>   vol. 21, 1973

Technology that tops nature. . . . Extracting or re-creating the pure special individual flavors of popular foods is an exacting science. . . . Ask [our] technologists how you can top nature's taste secrets—profitably.
>   —advertisement, *Bakery and Production Management,* November 1972

Just in case you are interested in some newly flavored ice cream consider then Tuna fish ice cream, tomato sherbet, jalapeno pepper and Lox'n Bagels ice cream. These flavors for ice cream have actually been concocted by one of the leading ice cream manufacturers in this country. Other ice cream flavors were "here comes the fudge, prune giggle, raisin cane, sugar plum and grape Britain." Some of these flavors were developed for real, some for fun.

—*Dairy Research Digest,* February 1976

The business of fragrances and flavors is as much an art as it is a science. We wrote the book on taking nature further. Send for your copy, it's a masterpiece.

—advertisement, *Food Processing,* January 1976

Although a rich diversity of . . . synthetic flavoring agents can be manufactured, the resulting tastes, interesting and attractive though they may be, cannot so far be made identical with those derived from nature. This may not be important; it is nevertheless worth recording. . . . So far . . . it seems that the production by synthesis of a flavor the same as that of natural foodstuffs has defeated the ingenuity and knowledge of the chemist. This is because of the complexity of the mixture of components, some of which are present in extremely minute amounts, of which natural flavors are composed. Synthetic foods are, therefore, likely to taste different from natural ones; they may, however, be none the worse for that.

—Magnus Pyke, *Synthetic Food* (London: John Murray, 1970)

There are very few flavors we can't reproduce. We make flavors happen.

—advertisement, *Food Processing,* June 1975

Good news! In these days of dwindling supplies and high costs . . . our new COUNTERFEIT CHOCOLATE FLAVOR may answer a lot of production problems. It is imitation chocolate flavor but its robust taste and aroma are just like the real thing! . . . Try it. One taste, one whiff and you'll wonder how anything that good could be counterfeit.

—advertisement, *Food Technology,* November 1974

Writing on mushrooms in the *Commercial Grower* [England] of 17th June [1960], Robert Patterson said: "It is suggested by some people that synthetic compost does not produce as good-flavoured mushrooms as those from horse manure compost." It is possible that there is something in this statement, too. A public analyst recently stated that, in bread making, mechanization has stopped the sweat from the baker getting into the dough,

100

Creations by the Food Flavorists

and that approval had been given for the addition of a certain chemical, the basis of which was the aid of human sweat. So it is just possible that some of the equine extract will have to be added to synthetic composts to supply that desirable horsey flavour.

—*Members' Notes,* The Soil Association, October 1960

〉〉〉〉〉〉〉〉

Food processors recognize that when basic foods are converted into highly processed products, natural flavorings suffer severe deterioration or even complete destruction from severe heat treatment or other forms of processing. Traditionally, flavoring ingredients have been added in attempts to restore these flavor losses.

With the advent of many fabricated foods, flavoring ingredients assumed even greater importance. Flavors need to be created in order to approximate the characteristic taste of the foods they imitate or to mask undesirable flavors of some components. Food processors place major emphasis on the development of flavoring systems in order that consumers accept their products. Flavoring ingredients, now numbering some fifteen hundred in use in the United States, form the largest single category of food additives.

Formerly, most flavoring systems were based on natural flavoring materials from spices, herbs, essential oils, oleoresins, and extracts from fruit, meat, and cheese. By definition established by the FDA, such natural flavorings must be extracted totally from natural ingredients and cannot contain any adulterants, diluents, or additives.

With increased demands for flavoring systems, and more specialized ones tailored to highly processed and fabricated foods, two types of artificial flavorings were developed: imitation and synthetic. The imitations may contain both natural and synthesized components, whereas the synthetics are created solely from chemical intermediates in the laboratory.

At present, more than twenty American companies devote their skills exclusively to the production of flavorings, and some of the major food processors produce their own flavorings. The largest flavor producer sold over $200 million worth of flavorings in 1974.

101

Between 1972 and 1975, dollar sales for synthetic flavorings increased by 20 percent annually. In anticipation of ever-expanding markets for convenience and fabricated foods, the overall synthetic flavoring market is expected to enjoy a steady growth in the years ahead.

The development of highly sophisticated analytical tools has made it possible to create an enormous range of synthetic flavorings. Components of natural foods can be separated and identified by gas chromatography and mass spectrometry. Computers quickly identify the main components, while researchers devote time to study the more elusive, or even unsuspected, ones. Analysts can determine which of the components in a particular food—which may number in the hundreds, or even thousands—are the most crucial for its taste, then attempt to approximate them.

These techniques have permitted flavor specialists to create a range of products that go far beyond the basic ones like vanilla, chocolate, strawberry, peppermint, or lemon. There is not only bacon flavor but artificial smoky bacon, maple bacon, and mushroom bacon created for snack food use. There is not only strawberry flavor but jam-type strawberry flavor; imitation strawberry flavors for hard candy, gelatin, and dessert mixes; and imitation strawberry flavor oil solutions for frostings and icings. Synthetic tomato flavors include one with a fresh, crisp tomato flavor, another with a pulpy note, a third with a cooked-tomato character. Synthesized tomato-flavor oils are reported to "closely duplicate the flavor and fragrance of fresh ripe garden tomatoes," whereas others imitate green garden tomatoes. Onion flavors have been created to simulate a "french-fried onion flavor," "toasted onions," and "green onions," which in turn flavor roasted, blanched, slivered almonds.

A description of specialized flavoring products offered to the trade by one supplier includes one that "faithfully duplicates the hearty taste and aroma of oven-roasted beef," another that "provides a flavor profile similar to stewed chicken," one that "achieves the succulent flavor of roasted pork," another with "a stout flavor similar to beef extract," one with "smooth beefy flavor with a charcoal-broiled character," and another with a "charcoal-like taste." The company also offers an artificial mushroom flavor that "achieves a smooth mushroom flavor without mushrooms" and an artificial clam flavor that "brings a smooth clam flavor to chowders and bisques."

Another company proclaims, "Our chicken is chickener. Flavor is our business. . . . Today we are proud to announce a new, exclusive . . . chicken flavor which feels more like chicken than any other chicken flavor you have ever tasted . . . specially recommended for soups, sauces and gravies. . . . It tastes just like grandma's cooking."

Prepared soups, gravies, and casserole mixes may contain artificial flavors of beef, ham, chicken, or seafood. The art has been extended to the creation of highly specialized artificial flavors and aromas that simulate anchovy, tuna, frankfurter, pepperoni, Italian sausage, country sausage, pizza, chili, or tacos for use in snacks, prepared foods, and textured soy products. Artificial fat flavors are used in soup stock, gravy, and sauce.

New flavorings are being created at a dizzying pace. Food trade journals describe an artificial mayonnaise flavor that offers a substantial cost savings for salad dressings and sauces. Artificial coconut milk flavor may be used in chocolate-coated fondants for candy. One fluid ounce per hundred pounds of candy can completely replace the natural coconut that would be needed in the product. An imitation roasted peanut flavor is used to give a fresh-roasted peanut flavor to baked products, ice cream, filled candies, and coatings. A custard flavor imparts "old-fashioned" custard flavor to pies or puddings as well as to baked goods.

Combined flavors extend the range: "Lift your snack sales with this newest flavor creation . . . imitation tomato and bacon." That combination is suggested to processors for use in chips, dips, curls, croutons, pretzels, potato sticks, yogurt, and cheese snacks.

Flavorings have been devised to overcome certain problems resulting from food processing. Flavors may be added to make bread and other bakery products appear to retain an oven-fresh smell after the first twenty-four hours. Flavor enhancers have been created to give a fresh taste to canned fruits and vegetables. Flavor boosters may be used to add to the natural flavors of beef and other fresh red meats. The cooked taste of pasteurized milk, which characterizes condensed and canned milks, has been simulated in imitation condensed milk flavor. For flavoring purposes, the imitation flavor can replace real condensed milk in caramels, toffees, fondants, nougats, and candy bar centers. Similarly, an imitation boiled milk dry flavor restores a boiled-milk taste to instant pudding mixes which are prepared with uncooked cold milk, so that the result will have the flavor of cooked pudding.

Given a challenging incentive, food flavorists boast that they can reproduce almost any flavor or aroma. In the early 1960s, *Aerosol Age* reported, "A chemically-created 'sales scent' is enticing supermarket customers toward particular display counters. Odors that can be sprayed from cans include chocolate, vanilla, banana, lemon, licorice, fresh bread, tobacco, and hickory barbecue. They're counted on to sharpen shoppers' buying impulses."

A California restaurateur had a modernized place complete with microwave ovens, but it lacked the traditional kitchen aromas that entice patrons. Flavorists responded with aerosol scents tailored to the restaurant's needs: baked ham and chemically fabricated Dutch apple cake.

Flavor Potentiators

Flavor potentiators can alter flavor in food without adding any flavors of their own. Under certain conditions, many ingredients act as flavor potentiators, including salt, sugar, some spices, maltol, nucleotides, and monosodium glutamate. Flavor potentiators may be used to intensify flavors, to smooth out or round off the rough edges—especially some artificial flavors—and to decrease or mask other flavors. The newer sources of protein for fabricated foods—soy, fish protein concentrate, cottonseed cake, petroleum-based protein—all present strong, rather unpleasant flavors.

The use of flavor potentiators allows food processors to reduce the amounts of ingredients that would ordinarily provide the flavor, so the nutrition in the finished product is diminished.

Potentiators are USDA approved and can be used in school lunch food. A line of oil-soluble, water-soluble, or spray-dried potentiators are applied to a wide range of food.

Creations by the Food Flavorists

Several new potentiators are characterized by unique qualities. One has a time factor, which produces an early flavor impact and also eliminates any lingering and undesirable aftertaste. Another has a fullness effect, which produces a full-bodied and creamy mouthfeel for low-calorie food. One has a wetness factor, which produces an impression of wetness in the mouth when dry or sticky foods are eaten.

Monosodium glutamate has received much publicity as a potentiator that causes many allergic reactions in sensitive individuals. Another potentiator, the purine nucleotides, is less well known by the consuming public. W. O. Caster, from the University of Georgia, reported that purine nucleotides are "sheer poison" for sufferers of gout. Caster charged that the nucleotides as flavor potentiators can lead not only to severe pain but to a whole variety of serious clinical problems. There is no restriction on their use in foods served in public eating places, nor any requirement for warnings on menus.

Many Flavoring Ingredients Now Replace Food Components

A "new natural egg flavor" is offered to the trade for use in "no-egg foods" such as scrambled eggs, mayonnaise, French toast, eggnog, pancake batter, cookies, batter for deep-frying, cake, and salad dressing. The flavor, made from "all natural ingredients," is described as having "a completely convincing egg flavor."

The flavor of creamery butter, prized by food processors, restaurateurs, and homemakers, is difficult to simulate, since more than a hundred volatile compounds have been identified in butter. But substantial new information about the flavor of butter has been uncovered during the last few decades, much of it resulting from attempts to simulate butter flavor in other products.

Imitation butter flavor, available in various forms, is advertised as having the flavor and aroma of pure butter and is said to "mellow or round out other flavors to a more pleasing total profile." Some imitation butter flavor is especially designed to withstand the high temperatures in making extruded food products. Imitation butter flavor is sold to food processors with a promise of considerable cost savings. One pound of imitation butter flavor is reported by one manufacturer to develop the flavor strength of approximately thirty-two pounds of creamery butter. Stored properly, such imitation butter flavor keeps fresh for six months and will not turn rancid in the container or in the finished product for many months. Imitation butter flavor is suggested for use in candies, snacks, butterscotch syrup, caramel corn, baked goods, frozen foods, sauces, soups, and canned foods.

A line of dry concentrated butter flavorants, derived from dairy products plus artificial flavorings, offers food developers "an opportunity to reproduce the flavor of fresh butter in foods where low fat or no fat is intended." The product is also available as a pump solution for poultry. In this case, the flavorant is injected into the fowl and becomes dispersed throughout the vascular system.

For years, bakers have resorted to imitation butter flavor as an economy measure. An oil-based flavoring product that imparts a buttery aroma is suggested for use with split-top bread loaves. Artificial cultured butter flavors are recommended for thrift in butter cream confections, butter sauce for brown-and-serve rolls, fresh bread, and other baked goods.

Restaurants and other institutions are offered liquid artificial butter-flavored dressing oil to replace more costly butter and the product is recommended for use as a "dressing, topping or seasoning ingredient on vegetables, potatoes, meats, poultry, seafoods, fish, egg dishes, sauces, gravies, waffles, pancakes, or as glazes for bread and rolls."

"Butter-flavored nuggets" consisting of butter fortified with artificial butter flavor and stabilized with a hydrogenated animal fat system are suggested to improve the texture, flavor, and appearance of preformed hamburger patties. The nuggets, irregularly shaped in sizes up to one quarter inch, do not require refrigeration, are highly resistant to rancidity, are easily dispersible, have the color and flavor typical of the product they represent, and when cooked with foods release a butter aroma.

With increasing acceptance by consumers of imitation foods, imitation butter flavor products are now available in retail stores. These items are suggested for use in preparing homemade biscuits, dumplings, pastry dough, cookies, cakes, breads, and waffles, and in replacing butter with rice, noodles, frozen vegetables, white sauce, and popping corn. The ingredients of one imitation butter flavor product are diacetyl and other ketones, butyric acid and other organic acids, vanillin and other aldehydes, ethyl butyrate and other esters, U.S. certified food color, propylene glycol, and water.

These days one wonders just how many real vegetables are used in prepared soups, casseroles, and other dishes. Various artificial vegetable flavors are employed as enhancers. Processors admit that some vegetable crops, bred for high yields, lack flavor, and artificial vegetable flavors are used in order to "add impact." There is also candid acknowledgment that "the greater use of chemical fertilizers, herbicides and pesticides and the development of new hybrid strains of grains, vegetables, fruits and new breeds of animals and poultry, assuredly increase the yields. Equally, they often decrease natural flavor levels and change textures which also affect flavor." Artificial vegetable flavors are also used to restore flavors lost in extreme processing. Potato flavor may, for example, be used to replace the lost "flavor notes" in dehydrated potatoes and to add "backup flavor" in extruded and baked snacks, soups, and similar products. Other artificial vegetable flavors that enhance—and may to some extent replace—real vegetables include cucumber, tomato, mushroom, string and lima bean, artichoke, pea, asparagus, corn, scallion, spinach, celery, lettuce, cabbage, bell pepper, onion, garlic, carrot, and soup greens.

Food processors favor artificial onion and garlic because they "impart consistent, stable flavor levels to food products while eliminating bacteriological and 'hot spot' problems." Such products, in dry form, may be prepared from encapsulated artificial oils of onion or garlic, which release the flavor instantly when blended with moist foods. Encapsulation helps to reduce the volatility of flavor and aroma and controls the release time of the flavor. It assures a shelf life of at least one year, contrasted to that of fresh onion or garlic, which may sprout or mold. The increased costs of onion and garlic products, along with the need to maintain handling and processing equipment for use with fresh products, all combine to make the use of such artificial vegetable flavorings attractive to processors. They are used increasingly as flavor sources or enhancers in soups, salads, dressings, sauces, casseroles, snacks, and dips.

Imitation and Synthetic Fruit Flavors

Shortly after a television program presented a spoof of fabricated foods—including a frozen lemon cream pie that contained no real lemons, no eggs, and no cream—the pie processor sent a memo to all personnel in its marketing system, analyzing the program in detail. The company was particularly sensitive to the publicity given the fact that the pie was lemonless. Not so, the company reported. Although it is true that artificial lemon flavor is used in the pie, a small amount of real lemon juice is present in the flavoring. Consequently, the company felt it was justified in listing as ingredients on the pie package, "lemon juice and other natural and artificial flavors."

Fruit flavorists boast that they have the "capability of making totally synthetic fruit flavors with the nuances like the freshness of orange juice or the tinniness of canned pineapple." Spray-dried artificial orange juice flavor is advertised as providing "fresh squeezed orange juice character."

"Most imitation pineapple flavor tastes like it," advertised a flavoring company aiming at food and beverage processors. "Only [our product] tastes like pineapple at its peak of flavor. The sweetness and full bouquet of pineapple sun-ripened in its natural environment. . . . That's the flavor of [our] imitation pineapple."

A wide variety of imitation and synthetic fruit and berry flavors is used in many processed foods and beverages. The range is far beyond consumer awareness, extending to flavorings that imitate fig, apple, raisin, honeydew melon, watermelon, and scores of other fruits.

From the processor's viewpoint, natural fruit flavors, regardless of their quality, are expensive, are in limited or irregular supply, and are variable in flavor and color. Synthetic fruit flavors, whatever their quality, are cheaper, are more stable in price, offer an adequate and dependable supply, and are uniform, batch after batch. Fruit flavor experts contend that as the use of natural fruit flavors declined, better imitation and synthetic flavors were created by improved technology. In some instances, products were developed that could not have been marketed without the use of imitations and synthetics. Fruit flavorists admit, frankly, that synthetic flavorings are not always equivalent in quality to natural ones, but they view the situation as a trade-off. The synthetics are judged adequate and contribute considerable cost savings.

However, there are many indications that discriminating consumers prefer natural flavorings and are willing to pay premium prices for quality products. A number of flavoring and fragrance companies still supply natural fruit flavorings, and in recent years many food and beverage companies have introduced products containing natural flavorings. The trend has been demonstrated especially with quality ice creams. A "natural" ice cream is made by 36 percent of the ice cream plants in the United States, with 91 percent of those reporting that their suppliers of "natural" ingredients guarantee their naturalness.

Replacers for Flavoring Seeds

Flavoring seeds, such as caraway, now have their replacers. "Caraway seed shortages facing you? High prices baffling you?" asks a flavorist in the trade. "If so, why not try . . . caraway flavor. Our product acts as a replacement for the flavor and aroma of caraway seeds. Use with or without seeds. If you want the old familiar appearance use 50 percent or less of the seed. . . ."

Artificial Spices and Recreated Spice Particles

As prices increased, companies selling traditional spices developed artificial spices as substitutes. One large spice company has developed artificial spices for about half of its line of fifty spices, including popular ones such as pepper, ginger, cinnamon, and nutmeg.

Spice grains may now be products formulated from extracts of real spice, with added cereal solids, emulsifiers, and other ingredients. According to the creators, the resulting spice particles have an "exceptionally natural appearance." Black pepper has black, gray, or white particles. Cinnamon has a mottled red or brown appearance. Other spice particles look like the natural products.

Since these spice replacers are manufactured, they can be altered to provide specific characteristics that a food processor may want in a product. The strength, color, and aroma can be precisely controlled. Such tailored products are particularly useful to bakers for mixes that require certain color characteristics in cinnamon or cloves; to snack manufacturers who may want extra color in chili powders; to sauce makers who may need specks of pepper; to pickle packers who may require extra-strength flavor for their products; and to a host of frozen food packers, canners, and other food processors who may find special applications for their products. The re-created spice particles are encapsulated so that the flavor lasts longer in "instant" products.

Vanillin Replaces Vanilla

Although the synthetic flavor and fragrance industry in America is more than fifty years old, its first major penetration into the American food supply was in the 1960s, when vanillin replaced vanilla to a large degree. Real vanilla flavoring, reportedly introduced into America by Jefferson, has had a venerable history of use as an extract, flavoring powder, and concentrate. By the 1960s, prolonged shortages of vanilla bean supplies led to widespread use of vanillin, which is produced inexpensively as a by-product of wood pulp. Vanillin became commonly used in carbonated beverages, ice cream, fruit drinks, bakery products, and many other processed foods, and also by homemakers. Vanillin's lower cost and steady supply were attractive.

"There hasn't been enough real vanilla available in the world for a century to flavor ice cream, much less other things," reported a spokesman for a flavor and fragrance company. "Most good vanillas are mixtures of the natural and the artificial. The cheap ones are all artificial."

In 1960, when the FDA proposed federal standards for the proportion of vanilla beans required in vanilla extract, food industry representatives proposed that the counterparts of pure vanilla extracts and flavorings be declared on food labels by phrases such as "contains vanillin, an artificial flavoring" or "vanillin, an artificial flavoring, added."

Three years later, when federal standards were adopted for frozen desserts (ice cream, ice milk, sherbet, and frozen custard), label requirements were established:

Vanilla Ice Cream: A product labeled in this manner must be flavored with 100 percent real vanilla flavoring from pure extracts or beans.

Vanilla-flavored Ice Cream: The flavoring in a product labeled in this manner may consist of up to 49 percent artificial vanillin. The word *flavored* must be present, but it does not have to appear larger than half the size of the word *vanilla.*

Artificially flavored Vanilla Ice Cream: The flavoring in a product labeled in this manner may be anywhere from 50 to 100 percent artificial vanillin.

More recently, the FDA requested that the food industry carry similar flavor declarations on the labels of other products. *Vanilla Pudding* or *Vanilla Cookies* would designate that the flavoring was entirely from vanilla, whereas *Vanilla flavored Pudding* or *Vanilla flavored Cookies* would indicate the presence of artificial vanillin.

The vanilla growers of Madagascar, the world's main source of vanilla, have been conducting a vigorous campaign to have food processors use an international symbol called the Vanillamark on food products that contain 100 percent vanilla flavoring and to urge consumers to look for products bearing that logo. The growers, through their trade organization, contend that "real, natural flavors, in general—and particularly vanilla—are more subtle. . . . [They] are more expensive than artificial ones and therefore, manufacturers reserve them for use in their higher quality products—ones commanding a better price. . . . They enhance good, fresh ingredients, but they will not mask out any off-flavors from other ingredients . . . meaning that the products in which they're used automatically require better ingredients overall."

As a result of this campaign, the demand for natural vanilla in cookies, ice cream, and other products increased about 16 percent from 1973 to 1975. Industry officials claim that the increased use stems partly from the growing interest of many consumers in natural foods, free of artificial ingredients. Also, government labeling regulations now make it easier to identify natural ingredients, and some food processors feel that the extra cost of real vanilla and other natural flavorings is a worthwhile sales tool.

Pure Maple Syrup versus Pancake and Waffle Syrups

Through the years, production of maple syrup in the United States has gradually dwindled. Around 1900 more than 5 million gallons were produced annually. By 1972, production had declined to only 1 million gallons annually, yet at the same time, the demand for it increased. Market studies showed that some 18 million American households used pure maple products (mainly maple syrup, but some maple sugar, too) during 1971. The use, however, was infrequent, and half of the consumers who thought they were using pure maple syrup were actually using a blend of less expensive ingredients such as corn syrup, water, sugar, and artificial coloring and flavoring.

One of the well-known blends was introduced on the American market as early as the 1880s, intended as an economical substitute for pure maple syrup. At the beginning, the product contained 45 percent maple syrup. Over the years, as maple syrup became more expensive, the percentage of maple syrup was gradually reduced.

Another well-known blend, a relative newcomer introduced in 1966, contained 15 percent maple syrup. As the percentage of maple syrup was gradually reduced, it finally became so low that the company concluded that the amount of maple syrup in the blend contributed scarcely any flavor to the product and could be eliminated completely. A company representative said, "Americans have come to like—even prefer—the taste and consistency of imitation syrup. Our consumer tests showed that many people like the artificial stuff better than the real thing. . . . Over the years, they've gotten used to it."

Other products of maple-flavored blends traditionally contained 15 percent maple syrup. Scarcities and rising prices for maple syrups led many companies to cut their blends to 10.5 percent. At first the reductions were made in states that didn't require that percentages be stated on labels, with the 15 percent being maintained in states with label requirements. The percentage was gradually cut to 10.5 in all states to achieve uniformity and was later lowered further, with some reaching 3 percent. Official standards did nothing to encourage quality for these blends. Gresham's Law prevails: bad quality drives out good. Syrup blends need contain only as little as 2 percent maple syrup.

The market studies cited earlier demonstrated that many people purchase maple-flavored blends with the erroneous belief that they are buying pure maple syrup. Blends that abound on grocery shelves reinforce the notion. The pictures on labels depict farmers collecting sap from the maple sugarbush or with ox-drawn sleds, or settings in rural kitchens or rustic cabins.

Consumer deception is added to consumer confusion. In 1975 a coffee shop in New York City was cited for a violation by the Department of Consumer Affairs for listing "real maple syrup" on the menu but actually serving a blend. Commissioner Elinor Guggenheimer reported, "Nothing in this syrup has ever been close to a maple tree. If maple syrup is listed on the menu, then it should be served—not a combination of chemicals, artificial flavorings and colorings."

Synthetic Chocolate Flavoring: "We Fooled Mother Nature"

After Cortez discovered chocolate in the New World, Spain maintained rigid control of the cacao trade. Although chocolate was introduced into Europe, it remained a luxury item that only royalty could afford. The price remained prohibitively high until British and Dutch explorers, finding new cacao sources, broke the Spanish monopoly. Within the next two centuries chocolate became popular throughout Europe as well as the New World.

In modern times, chocolate has been a favorite flavor in the United States, with an annual domestic market exceeding $1 billion. But history seems to have repeated itself. By the mid-1970s, the price of cacao beans soared and once again chocolate became a luxury item.

Ghana, which produces some 40 percent of the world's cacao bean crop, suffered drought and severe crop failures. During the same period, worldwide demand for cacao beans increased. The sixteen main cacao-producing countries in Africa, South America, and Asia saw the success of price manipulations by oil-producing countries and considered cutting their production as a means of gaining political clout and reaping higher profits. The various factors combined, and by 1973 the wholesale price of chocolate nearly doubled, then tripled.

Candy manufacturers using chocolate coatings were especially affected. Traditionally, two basic ingredients of real chocolate have been used in compound coatings (also called confectioners' or confectionery coatings): the "chocolate liquor" squeezed from the cocoa bean, and the cocoa powder residue. Cocoa butter is derived from further processing of the chocolate liquor.

By the mid-1970s, three out of five leading chocolate candy bar manufacturers had begun to market candy bars in which cocoa butter was completely replaced by hydrogenated oils from soybeans, coconuts, or palm kernels. They had developed a confectionery flavor system that combined artificial flavor and a bulking agent to replace all or any portion of the chocolate liquor or cocoa and sugar normally used. Although this flavor system was intended primarily as a candy coating, its use could be extended to fillings or in solid bars. Various reports promised manufacturers savings of from 17 to 27 percent over the cost of real chocolate flavoring.

In addition, chocolate flavoring extenders were developed, designed to replace some chocolate and consisting of flavoring materials, thickeners, starches, hydrogenated fats, and colorings. Such

114

extenders have helped candymakers reduce their costs substantially by replacing 25 to 50 percent of the cocoa butter in confectionery products, especially the confectioners' coating used on candy bars.

Imitation chocolate flavors, cocoa replacement flavors (imitation and natural), chocolate enhancers, and extenders are expected to be used increasingly, widely applied for candies, baked goods, frozen desserts, low-calorie foods, milk drinks, carbonated beverages, and cordials. With baking chocolate costing about two dollars a pound by the mid-1970s, items such as chocolate-flavored frosted doughnuts contained no chocolate at all. Chocolate-flavored baking chips were introduced in which cocoa butter was replaced by hydrogenated vegetable oils. Some state consumer protection agencies issued warnings urging consumers to read labels carefully in order to distinguish between chocolate baking chips, which were made with real chocolate, and chocolate-flavored baking chips, which were as much as twenty cents cheaper per twelve-ounce package. The agencies cautioned that though such packages might be properly labeled, consumers could be confused by added descriptions such as "new!" or "nu-style."

One new artificial chocolate was marketed with claims that "our newest invention will go unnoticed. It's the true taste of chocolate made entirely without chocolate . . . now manufacturers can produce chocolate-flavored products of all kinds and maintain a consistent profit margin . . . our artificial chocolate actually tastes like the real thing, and nothing else."

Food technologists themselves have been less than enthusiastic: "Unfortunately these 'chocolate' coatings, composed principally of vegetable fat, cocoa, sugar and vanillin, are poor replicas of real chocolate." Most of the new compound coatings have a higher melting temperature than real chocolate, and they fail to melt in the mouth in the same way.

One food writer described the chocolateless chocolates as "cheap tasting imposters." Consumer surveys have also reflected disenchantment. These products have been variously described as "waxy to the palate," "tasteless to the tongue," "gritty," "flat," "cloyingly sweet," or "a sourish super-sweet ersatz, with an acrid aftertaste and an aroma not unlike cheaply perfumed face powder."

115

Food Flavorings: How Safe?

Some fifteen hundred food flavoring agents, both artificial and natural, are countenanced by American authorities. The French government, on the other hand, permits the use of only seven synthetic flavorings. The French Academy of Medicine officially opposes all practices that lead to the introduction of foreign substances into foodstuffs, even if they are reputedly harmless. The German government restricts to eleven the synthetic flavorings "not found in nature." The philosophy guiding the German policy is that foods should preferably be flavored with substances derived from natural sources; when synthetic compounds are used, after being tested and found harmless, the products containing them must bear a declaration stating that they contain an "unnatural" flavoring agent.

Food flavoring regulations differ from country to country, and substances prohibited in some are permitted in others. For example, though pennyroyal is banned as a food flavoring in Great Britain, West Germany, and Yugoslavia, two types of pennyroyal are permitted in the United States. Quassia wood, forbidden in West Germany and Yugoslavia, may be used in the United States.

The Food Standards Committee's Report on Flavoring Agents, issued in Great Britain in 1965, listed sixteen flavoring agents as having been shown to be harmful or suspected of being harmful: coumarin, tonka bean, safrole, sassafras oil, dihydrosafrole, isosafrole, agaric acid, nitrobenzene, dulcamara, pennyroyal oil, oil of tansy, rue oil, birch tar oil, cade oil, volatile bitter almond oil, and male fern. The committee recommended that these flavoring agents be prohibited for use in food.

Several food flavoring agents used in the United States in the past were ultimately banned:

Coumarin: an artificial flavoring used in synthetic vanilla, chocolate, and other confections for nearly seventy-five years, was found to cause liver damage in rats and dogs. Its natural source, the tonka bean, was banned, along with coumarin, in 1954.

Dulcin (ethyoxyphenylurea): an artificial sweetener 250 times sweeter than sugar, was synthesized in 1883 and used for more than

116

half a century in foods. When tested, dulcin was found to cause liver cancer in rats. It was banned in 1950.

Oil of calamus: a flavoring derived from the calamus plant root, was considered "Generally Recognized As Safe" (GRAS), but later was shown to be carcinogenic. It caused intestinal cancer and damaged the liver and heart in rats. The use of oil of calamus in food was banned in 1968.

Safrole: a flavoring substance extracted from sassafras bark, was used in the production of root beer and similar beverages. When tested, safrole was found to cause liver cancer in rats and in dogs, so was banned from food use in 1960.

The more recent, much-publicized investigations of the two synthetic sweetening agents, cyclamates and saccharin, are still unresolved. The cyclamates, synthetic sweeteners approved for special dietary purposes in 1950, were considered GRAS in 1958. By 1970 cyclamates were found to induce bladder cancer in test animals and were banned. Although further investigations failed to determine the cancer-inducing properties of these sweeteners decisively, a new set of problems was raised with the finding that cyclamates induced both noncancerous and cancerous tumors in animals, stunted growth, interfered with reproduction, produced testicular atrophy, and elevated blood pressure. In humans, cyclamates damaged chromosomes when ingested at doses approximating ordinary use.

Similarly, many questions about saccharin remain unresolved. Although this synthetic sweetener has been consumed for nearly a century, long-term usage does not necessarily assure safety. By modern standards, early safety tests for saccharin were notoriously crude and unreliable, having been performed with impure substances or with unhealthy animals. Numerous medical reports have described adverse effects of saccharin on humans, including allergies, cardiac arrhythms, and gastrointestinal upset. Saccharin was an acknowledged enzyme inhibitor. In animal studies, saccharin was found to concentrate in certain kidney or bladder tissues. It was suspected as a mutagen, a teratogen, and a cocarcinogen (a substance that may promote cancer by acting synergistically with one or more substances that are carcinogenic).

On 7 March 1977 the Canadian government released findings that were reported to demonstrate unequivocally that saccharin induced bladder cancer in rats, with a sharply increased incidence shown in the second generation of animals. The Canadian government banned saccharin for general use but announced that it would make it available as a drug for diabetics. By extrapolating the Canadian data, it was estimated that 4 out of every 10,000 people who drank only one twelve-ounce can of a diet soda daily over a lifetime would run the risk of developing bladder cancer. "This number times 215 million Americans would be a public health disaster," stated Sherwin Gardner, acting commissioner of the FDA. Gardner's position was reinforced by Dr. Guy R. Newell, Jr., acting director of the National Cancer Institute, who reported that projections from the Canadian results indicated a possible additional 600 to 700 cases of bladder cancer among Americans each year. On 9 March 1977 the FDA proposed a ban on saccharin. In June 1977, a Canadian study provided the first direct evidence linking saccharin consumption to human cancer. Although under the Delaney Clause the FDA was obligated to restrict or prohibit further use of saccharin, the agency was hamstrung by Congress. Numerous bills resulted in weak action, with only a suggested warning label, and an eighteen-month moratorium precluded any prompt action.

Food flavoring agents include compounds with a wide variety of chemical structures and mixtures of variable compositions. In recognition that toxicity data were woefully lacking for many food flavorings, the FDA conducted studies with a large number of these substances for acute oral toxicity in rats, guinea pigs, and mice. (Acute, as contrasted with chronic, toxicity refers to harmful effects that occur within a short period of time, instead of lasting or lingering injuries to health.) Numerous adverse effects were observed, including rough fur, scrawny appearance, diarrhea, soft stools, bloody urine, intestinal irritation, hemorrhaging, anatomic or functional manifestations of disease conditions in the liver, lacrimation, salivation, gasping, respiratory failure, tremors, depression, ataxia, coma, and even death.

In later studies, the FDA tested certain food flavorings and related compounds for subacute and chronic toxicity in rats. Many of the substances caused retarded growth, increased mortality, damage to organs such as the liver, kidneys, and heart; stomach ulceration; bone changes, and malignant tumors. Since the FDA has maintained a policy of not publishing what the agency has learned about the toxicity of food additives, reports of its studies with flavoring materials are not apt to be found in technical journals or in government publications. There is a great scarcity of trustworthy toxicological data for many (perhaps most) of the food flavoring agents used by food processors. These compounds, regarded as valuable trade secrets, are treated by the FDA as privileged materials. Despite the FDA's proposal to expand public disclosure of information previously held as confidential, the agency considers such information as food flavoring safety test data to be trade secrets and therefore exempt from the Freedom of Information Act.

Dr. Jacqueline Verrett, a research chemist with the FDA, has publicly cited two officially sanctioned synthetic flavorings that are possibly dangerous. A synthetic pineapple flavor may cause liver damage. Gamma valerolactone, a flavoring used in ice cream, candy, and baked goods, is suspected of being carcinogenic.

Many essential oils used as food flavorings have enjoyed GRAS status. However, in tests for chronic toxicity some were shown to be mildly irritating to the mucous membrane of the mouth and the digestive tract. Their ingestion in large amounts irritated the kidneys, bladder, and urethra. In the presence of a preexisting inflammatory condition of the urinary tract, small doses of ingested essential oils appeared to worsen the condition. Essential oils applied to mouse skins produced moderate to marked skin hyperplasia (an abnormal increase in the number of normal cells in ordinary arrangements in a tissue). In some cases, areas of necrosis (death of tissues) with ulceration, oozing or weeping of fluid, and crusting developed. Some essential oils were found to promote skin tumors. Benign warts and malignant skin tumors appeared in some animals that had first received a treatment with a known cancer-inciting substance and had then repeated skin treatments with an essential oil.

Food flavoring agents, according to Dr. Jean Mayer, are "one of the areas of greatest toxicological uncertainty at present."

12. Creations by the Food Colorists

For over 2,000 years a common purpose of adding [food] color was to defraud the consumer or disguise adulteration.
—*Food Colors* (Washington, D.C.: National Academy of Sciences, 1971)

The more sophisticated we become in our food habits and the more dependent we become on stored and packaged food, the greater the tendency to color our food.
—Professor E. Boyland, London *Times,* 13 February 1967

As long as the consumer wants food colors comparable to the rich yellow of egg yolk, the bright red of ripe tomato or the appetizing red of cooked lobster, food manufacturers must continue to use colorings with the best available technology and within the limits of government regulations.
—spokesman, Institute of Food Technologists, Anaheim, California, 1976

How our color laboratory can help you: service to customers is an important part of [our] business. . . . Does a competitive product have more eye appeal? Let us make recommendations.
—*Certified Food Colors* (St. Louis: Warner Jenkinson Company, undated)

Purpose is outside the boundaries of science. In Britain, the confectioner's customers like their cakes coloured. The chemist, understanding the rela-

Creations by the Food Colorists

tionship between colours . . . can readily synthesize any colour the food manufacturer wants; yellow to remind the eater of eggs, brown to simulate chocolate, or red or blue or green just for fun.
—Magnus Pyke, *The Boundaries of Science* (Middlesex, England: Pelican-Penguin, 1961)

The addition of a dye without nutritive value to food must always be a subject for concern. . . . . The practice of dyeing food is an old one but that does not make it any more respectable. Dyeing always raises the possibility of concealment, of making something look better than it really is. It would seem that the smaller the list of colors permitted and the more thoroughly they are investigated, the better. Why, indeed, does our food have to be dyed?
—Annotation on dyed food, *British Medical Journal,* 22 April 1961

Cosmetic effect is most important and it gives the competitive edge to those foods with the most appealing color. People don't want gray-colored hot dogs and sausages, and gray is their natural color. Also, how would you distinguish different flavors in gelatins all the same color?
—Earl M. Handing, Warner Jenkinson Co., 36th annual meeting, Institute of Food Technologists, Anaheim, California, 1976

Dyes in food are valueless: neither the nutritional value nor the keeping quality is enhanced. Food manufacturers claim that the consumer demands bright colours: but really the supply has created the demand, and even granted that the demand is now ineradicable the prohibition of synthetic dyes in foods would rapidly lead to the substitution of safe, if inconvenient, natural colors, such as carotene in margarine; or cause jam, tinned peas, etc. to be made from fresh unfaded produce, while kippers and haddocks would be smoked and not dyed.
—Franklin Bicknell, M.D., *Chemicals in Food & in Farm Produce: Their Harmful Effects* (London: Faber & Faber, 1960)

Mother nature is often more bland than many persons would wish. For example, one maker of "all natural" ice cream points out that many customers complain about there not being any green in its chocolate chip mint flavor. But pure mint is clear, not green; hence the "natural" product doesn't

121

possess the "sales appeal" of something a bit less natural and a bit more colorful. Ice cream with real strawberries is pale pink, and real vanilla beans, if ground up for use in pure form, are dark. Recognizing this preference (and, some say, stimulating it as well), food manufacturers use additives to enhance color.

—Timothy Larkin, special assistant to the commissioner of Food and Drugs, FDA, "Exploring Food Additives," *FDA Consumer,* June 1976

The coloring of food with coal tar dyes is a common manufacturing practice. However, the use of these nonnutritive food additives can only be justified on questionable aesthetic and psychological grounds.

—R. A. Chapman, "Nutritional Aspects of the Use of Food Colors," *Federation Proceedings,* 1961

We cannot accept the contention that, because "coal tar" colors have been used in foods for many years without giving rise to complaint of illness, they are, therefore, harmless substances. Such negative evidence in our view merely illustrates that in the amounts customarily used in food the colors are not acutely toxic, but gives no certain indication of any possible chronic effects. Any chronic effects would be insidious and it would be difficult if not impossible to attribute them with certainty to the consumption of food containing colouring matter.

—Summary of conclusions, *Food Standards Committee Report on Colouring Matters, Recommendations Relating to the Use of Colouring Matters in Food* (London: Her Majesty's Stationery Office, 1954)

>>>>>>>>>

The coloring of foods and beverages to adulterate or disguise blemishes or enhance the appearance of products is an ancient art. Through the centuries, color pigments have been derived from such plants as beets, saffron, annatto, paprika, and turmeric; from such insects as cochineal, and from such minerals as titanium dioxide and iron oxides. Pliny described the process of coloring wine between 200 and 300 B.C. A Parisian edict in 1396 forbade the coloring of butter.

Creations by the Food Colorists

By the early nineteenth century the coloring of foods and beverages was widely practiced and virtually unregulated. Many toxic substances were used: vermilion (mercuric sulfide) or red lead were used to color cheese rinds, copper arsenite or lead chromate to color tea leaves, and lead and copper salts to color candies. There are recorded accounts of death from eating pickles colored green with poisonous copper sulfate and from pudding colored green with poisonous copper arsenite.

By the middle of the nineteenth century food processors displayed keen interest in dyes being synthesized and used industrially. Pigments from natural sources deteriorate with exposure to light or air, with temperature extremes, or through interactions with other components in the product. Processes and procedures used in preparing foods for preservation or more convenience may alter them both chemically and physically and adversely affect the color. Processors regarded synthetic dyes as far superior to vegetable and mineral colorings: better tinctorial value with a large range of hues, greater stability and uniformity, dependable supplies, and lower costs. By 1900 nearly seven hundred coal tar dyes were marketed and many were being used in foods and beverages.

Why Foods and Beverages Are Colored

Colorants can mask adulteration. By 1900 yellow coloring was added to milk in London to prevent detection of the practice of watering or skimming the milk. Adding the coloring became so commonplace that housewives came to accept yellow as the normal color and viewed uncolored milk with suspicion. This practice continued in England until 1925, when it was outlawed.

In the middle of the nineteenth century a chemist, Dr. Arthur Hill Hassall, analyzed forty-nine loaves of bread from various London shops. All of them contained alum, which can make musty gray flour appear like fresh good-quality flour.

The use of colorants to mask adulteration is not merely a practice of the past. In 1962 the Department of Markets in New York City charged some butchers with selling hamburger meat containing as much as 90 percent fat. The illegal addition of beef blood colored the product and disguised the fat in the meat. Industry representatives conferred with the commissioner of markets to seek permission to continue using beef blood as a colorant in the fat, claiming that they were "performing a service to the poorer people by making the meat more nutritious and better looking." The request was rejected and the practice denounced as adulteration and deception.

Sodium sulfite, an additive, can mask the smell of deteriorating meat and give it a fresh-meat redness. Such meat is injurious, especially if eaten rare. Sodium sulfite is a poison that destroys Vitamin B and is capable of causing considerable damage to the digestive system and other organs. Yet tested samples of ground beef purchased as ready-chopped hamburger, or sold at hotdog stands, cafeterias, and restaurants, frequently show adulteration with this chemical.

Hamburger meat served in restaurants often contains sodium nicotinate to preserve its bright red color. Although this chemical is illegal in some municipalities, thirty-seven states permit its use. Several outbreaks of poisoning have been traced to this additive.

In 1974, saithe (a fish known as *Seelach* in Germany) was seized in a New York City store by a representative of the Department of Consumer Affairs. The saithe was colored by a food dye to look like the more expensive Nova Scotia lox.

Colorants can mask blemishes. Testimony during congressional hearings on chemicals in food products in 1951 revealed how a colorant disguised the bruises in fruit used in making maraschino cherries. To prepare the cherries for the red dye, the original color is bleached out, which helps disguise any blemishes such as dark pigmentation caused by rot spots. After the cherries are dyed it is impossible to determine from their color that some parts of the fruit may be rotten, so the dye can conceal inferior, bruised fruit.

The principal use of colorants in modern times is, however, to make foods and beverages appear of better quality than they really are. Thus, whether food colorants come from traditional or synthetic sources, they don't serve consumer interests. Although traditional sources of colorants may not offer the hazards of synthetics, they, too, should be regarded as deceptive.

Food and beverage processors claim that impressions of foods are often expressed in terms of characteristic colors. There are preferences for colors associated with the maturity of fruits and vegetables, the richness in gravy, the brownness in baked goods. Products that lack the expected color value may be thought inferior. Thus, pink rather than red fruit may suggest lack of maturity; very light brown bakery goods, insufficient baking, or unnatural color, a sign of spoilage. Butter, cheese, egg noodles, and lemon-flavored cakes are expected to appear yellow; mint-flavored jelly, green, and orange drink, orange.

Governmental approval of the use of food colorants that results in consumer deception begins on the farm with the postharvest treatment of certain crops. The skins of last year's white potatoes may be dyed red to make them look like expensive new potatoes. As early as 1961, an FDA spokesman warned the potato industry that the agency was receiving numerous complaints from consumers about the practice of coloring potatoes. Consumers were not being informed of the presence of the coloring at the point of retail sale. It was assumed further that, had the complaining consumers known that the potatoes were colored, they would not have purchased them. The agency spokesman said that the practice warranted very careful consideration. He warned, "If we encounter any lot where we believe we can demonstrate that the color serves to make the potatoes look newer or appear to be better quality than they are, we will take legal action against the goods."

Actions were not taken. By 1964, the agency reported:

> We have considered the possibility that such colored potatoes might be deceptive and therefore illegal, but we have not encountered an instance where we felt we had evidence to establish in court that the colored potatoes were, in fact, deceptive. To prove this we would need clear-cut proof that potatoes of a lower quality were dyed and sold as of a higher quality, or a lower priced white potato, after being dyed, was sold at a high price as a red variety of potato. Although we sympathize with those who object to the penetration of the color into the flesh of the potato, the law contains no provision which would permit us to outlaw the dyeing of potatoes solely for that reason. Certainly, however, the consumer has the right to know when the potatoes are colored. Under the law, not only must the bags of potatoes

be labeled to show that they are colored, but when sold in bulk in the retail store, a sign calling attention to the presence of added color should be displayed.

The flaws in this argument should be apparent. The FDA does have authority under the Federal Food, Drug, and Cosmetic Act to outlaw the practice of dyeing potatoes. Section 402, b.3 clearly states, "Damage or inferiority in a food must not be concealed in any manner." The agency considers the dyeing of potatoes a very minor problem. As recently as 1973, the agency reaffirmed its stance that the dyeing of potatoes is not a violative practice. Although the agency gives lip service to the principle that consumers should be informed by signs calling attention to the presence of added color in the potatoes at the point of retail sale, how many consumers see any placard in the potato bin stating that the potatoes are dyed?

The skins of sweet potatoes, too, may be dyed to give them heightened color and make them appear like yams, which are more expensive. This deceptive practice was banned in Canada. In June 1968 the FDA ruled that "artificially red-dyed yellow varieties of sweet potatoes would be deemed adulterated" and their interstate shipment was banned. Why is the same principle not applied to red-dyed white potatoes?

Consumers are short-changed nutritionally by government regulations that tolerate the addition of colorants to butter to give it a more appealing and uniform hue, despite seasonal changes. Consumers have no way of knowing if the rich yellow coloring of the butter is the natural appearance of the product, from cream given by cows on vitamin- and mineral-rich summer grasses, or if the appearance of the product is the result of color added to the pale butter from cream given by cows fed on nutritionally inferior winter hay. Consumers ought to have the information. Butter produced from summer pasture feeding may be several times as high in fat-soluble Vitamins A and D as butter produced from animals fed on winter hay. This information is not required on butter labels, nor is it even necessary to declare the presence of added coloring in butter, cheese, or ice cream.

Many consumers associate the yellow skins of poultry and deep-colored egg yolks with high-quality foods. The FDA allows the use of four feed additives as pigmenters to increase the yellow color in poultry skins and egg yolks: dried algae meal, aztec marigold meal, an extract of marigold meal, and corn endosperm oil. These substances supply xanthophylls, substances structurally related to Vitamin A, which are metabolized into yellow pigmentation. These feed additives achieve the yellow coloring in the poultry skins and egg yolks formerly achieved with more expensive but nourishing corn and alfalfa. The yellow color pigmenters are very commonly fed to chickens raised to supply eggs to processors. Such eggs, known as breakers, are used for cake mixes and egg noodles, where a dark yellow coloring is favored. Consumers are short-changed nutritionally by such practices whenever foods appear to be of better quality than they are in reality.

Another illustration of governmental approval of food color use that deceives consumers is the use of caramel and yellow colorings in baked goods. Caramel makes the baked goods appear darker, like loaves made with whole-grain flours, and the yellow makes baked goods appear rich in eggs or butter. Public interest scientists, including the Center for Science in the Public Interest (CSPI), petitioned the FDA to ban such use of colors on the ground that "any bread manufacturer who adds artificial coloring to his product is in violation of the economic adulteration provisions of the Federal Food, Drug and Cosmetic Act." The agency rejected the petition, but stated that it would reconsider its position if proof could be given that manufacturers, by adding dark colorings to their bread, deceived the public, which believed that the loaves were more nutritious.

In response, the CSPI conducted a survey of shoppers in supermarkets in the Washington, D.C., area. Up to 43 percent of the people who bought dark bread with caramel coloring thought they were actually buying a nutritionally superior loaf. They were unaware of the fact that many dark-colored loaves of bread are merely disguised white bread. Similarly, many consumers assume that bakery products with yellow coloring are rich in eggs or butter.

The FDA has not responded to the CSPI's survey. Although the agency ignores the issue, bakers are clearly aware of the intended use of caramel coloring in breads. An advertisement by a molasses company displays two pictures of dark breads identical in appearance. The captions under the breads read "ARTIFICIAL COLOR, ingredients: rye flour, wheat flour, water, salt, yeast and artificial coloring; 100 PERCENT NATURAL COLOR, ingredients: rye flour, wheat flour, water, salt, yeast and molasses." Then the advertisement asks: "IF YOU WERE THE CONSUMER, WHICH WOULD *YOU* BUY?"

Caramel is also used in many soft drinks, cola and alcoholic beverages (Scotch and Irish whiskeys, rum, brandy, and wine), ice cream and ices, candy, baked goods, syrups, and meats to impart a rich brown color. Although the use of caramel coloring has long been assumed safe, in the 1960s the Joint FAO/WHO Expert Committee on Food Additives classified caramel under "colors for which virtually no toxicological data were available." In 1976, the official Danish Foodstuffs Institute, basing its action on a WHO recommendation, ordered Danish processors of cola drinks to reduce the amount of caramel added to their beverages. The amount of caramel used in these drinks was declared a health hazard, especially for children who might regularly consume such beverages. The hazard lies in the fact that a specially processed caramel known as ammonized caramel is used as the colorant. Ammonized caramel has been found to contain several levels of 4-methylidazole (imidazoles and pyrazines). When 4-methylimidazole was administered orally to rabbits, mice, and chicks, it produced convulsions, which illustrates that colorants—even traditional ones assumed harmless—are undesirable in the food supply.

Artificial coloring of food may also be achieved by means other than adding colorants. A 1937 USDA regulation requires that vine-ripened tomatoes must be larger than those picked prematurely and artificially colored by ethylene gas. This discriminatory regulation, formulated during the Depression when the United States was attempting to favor American agriculture, was intended to limit the supply of vine-ripened tomatoes, imported mainly from Mexico, and to help raise the price of ethylene-gassed tomatoes produced largely in Florida. The regulation, still in force, illustrates Gresham's Law: poor quality drives good quality from the marketplace. Consumer groups have protested, asserting that vine-ripened tomatoes have higher levels of Vitamins A and C, taste better, and have better texture than unripe ones that are merely gassed to redness. Even if the official policy is reversed, it will not necessarily result in the disappearance of gassed tomatoes from the marketplace. As long as the practice is allowed, some producers will continue to gas tomatoes so they can ship to market sooner with less spoilage.

For many years the citrus industry was allowed to use a food dye, FD&C Red No. 32, to mask the greenness of the fruit rinds. Industry spokesmen insisted that the practice was necessary for consumer acceptance. Tests finally revealed the toxic nature of the dye. The FDA delisted the dye, with a lenient provision that the citrus industry could continue temporarily using it until a substitute could be found. In 1959 FD&C Citrus Red No. 2, described as less toxic than FD&C Red No. 32, was substituted. Industry was given another period of grace to seek a harmless alternative but failed to find one and used the time to lobby and receive, repeatedly, liberal extensions of time. Artificially colored oranges continue to be sold to American consumers, though Canada, Great Britain, and other countries will not permit artificially colored oranges to enter their countries.

Formerly, it was mandatory for oranges to be stamped "color added." Even this measure was dropped. Industry spokesmen claimed that the practice was too costly and time consuming. Despite these claims, some citrus growers find it convenient and profitable to stamp individual pieces of citrus fruit with their brand name or variety.

Consumer education is preferable to the practice of dyeing citrus fruit. Boxes of fruit have been packed by some groves with the following information leaflet enclosed:

Green Is Beautiful

Don't let my greenish tint bother you—inside I'm fully ripe. You see, my outside color depends on the climate. When it's cold I turn yellow-orange—the way you're used to seeing me. But this year, our nights have been mostly very mild, so my tree has been busily putting more natural sugar into me instead of color into my skin. I'm not only naturally beautiful, I'm delicious!

Food and beverage processors, in a highly competitive marketplace, will not voluntarily discontinue the use of colorants. They will continue to enhance the appearance of their products, especially since governmental agencies fail to discourage the practice. If the entire citrus industry would agree to discontinue dyeing the fruit, consumers would learn to accept green-colored oranges as ripe. Both in the Caribbean and in Central American countries, the sale of oranges with a green color is customary and there is no public demand for a uniform orange color in the rind.

Food Dyes: How Safe?

As mentioned, by the turn of the twentieth century many coal tar dyes were being used in the food supply, but there was no information about their safety. Dr. Bernhard C. Hesse, a German dye expert, was given funds by the Bureau of Chemistry (then of the USDA but later to become the FDA under HEW) to investigate the safety of food dyes. Hesse approved seven dyes, with blends for intermediate shades, and these were in turn approved under the Pure Food and Drug Acts of 1906. Between 1916 and 1926, ten more dyes were approved. Despite the belief that all of these dyes were safe, most of the seventeen approved dyes were banned at later dates. It is obvious that food dyes have an especially poor safety record.

Coal tar dyes were made from the black, viscous fluid obtained as a by-product in the manufacture of illuminating gas from coal. Formerly, coal tar was the only technically important source of raw materials for the dyestuff industry. Now these raw materials are obtained from petroleum as well. The method of synthesizing purifies them to such an extent that their original source is unidentifiable. It is thus more common to refer to such colorants as synthetic organic dyes or simply as synthetic dyes. The chemical classes of synthetic dyes from which U.S. certified food colors are made are azo, triphenylmethane, and xanthene. Tests of examples of all three classes have shown potential harmfulness.

The toxicity of azo dyes depends on how they are metabolized in the body. Some merely pass through and are excreted in an unchanged form. Others break down into innocuous products and are excreted. Some are reduced to supposedly harmless substances by living organisms normally found in the gastrointestinal tract. But at times azo dyes are absorbed into and become bound to organs and tissues of the body. It is this action that can harm. It has long been recognized that certain azo colorings can induce liver cancer in rats, by the colorings becoming bound to the protein of the liver.

A number of azo colorings produce foreign particles (Heinz bodies) in the blood. Some have produced malignant tumors (sarcomas) at the site of repeated injection of the dyes under the skin of experimental animals. There is some disagreement among experts as to whether the production of sarcomas by this technique necessarily indicates that the substance may be a cause of cancer when the dye is consumed orally in foods and beverages.

On the basis of animal experiments, triphenylmethane dyes are judged to be potential human carcinogens. Xanthene dyes have demonstrated mutagenic qualities. Most synthetic dyes currently in use are of questionable safety. Many of them have provisional approval, which means that their safety has not been established. On the contrary, data exist that suggest potential hazards. The FDA's policy of allowing these dyes to remain in use for many years under provisional status when their safety has not been established exposes consumers to unnecessary risks. Food technologists are quite aware of the fact that these food dyes may ultimately be banned. This was candidly admitted by Dr. Howard Mattson, public information director of the Institute of Food Technologists, at its annual conference in June 1977. "There's a possibility in six, eight, ten years that there won't be any synthetic dyes."

The editor of the *British Medical Journal* posed the question, "Why, indeed, does our food have to be dyed?" In a subsequent issue of the journal Franklin Bicknell, M.D., responded, "The answer can only be that food should never be dyed, since it is virtually impossible to be sure of the safety of any of the . . . currently 'permitted' synthetic coal tar colors. . . . With such uncertainty about the cancer causing properties of almost every dye, all should be banned." To this Dr. Joshua Lederberg added, "People won't weep if they can't get pastel breakfast cereal."

13. Natural versus Synthetic Vitamins

The evidence that has accumulated over many years since we've identified nutrients makes it very obvious that a nutrient is a nutrient, or a vitamin is a vitamin is a vitamin.

—Ogden C. Johnson, Director of Nutrition and Consumer Sciences, FDA,
FDA Consumer, December 1973

The FDA says there is absolutely no difference between synthetic and natural vitamins. In fact, the agency has proposed a regulation that would forbid any vitamin [manufacturer] from claiming on its label or in advertising that its "natural" origins made it better than the same vitamin from synthetic sources.

—Brattleboro [Vermont] *Reformer,* 17 March 1975

In spite of the therapeutic value of synthetic vitamins in the treatment of certain diseases, one should not believe that the mere ingestion of the pure synthetic vitamins, even if identical in chemical structure to the natural vitamins, necessarily has the same effect on our bodies as the consumption of vitamins in combination with their natural concomitants. *The latter are not to be dismissed as "excess baggage" but must be thought of as the result of millions of years of evolution of optimal combinations.* [Emphasis added]

—Alexander Berglas, *Cancer: Nature, Cause, and Cure* (Paris: Institute Pasteur, 1957)

Natural versus Synthetic Vitamins

On the whole, we can trust nature further than the chemist and his synthetic vitamins. Recently, Professor J. C. Drummond, the scientific advisor to the British Ministry of Food, voiced his reluctance to put the dietary safety of a nation on synthetic vitamins, as a long range policy. He thinks we must and should provide the natural vitamins in the natural foods. I stand on that platform, until we know a great deal more than we do today about foods and human nutrition.

—Dr. Anton J. Carlson, Professor of Physiology, Chicago University, *Science,* 30 April 1943 and 7 May 1943

Wherever possible, the nutrients required should be obtained from natural foods rather than synthetic preparations. This is particularly true of the vitamins. Synthetic vitamin concentrates, tablets and pills may not contain all of the known nutrients, either as such or in optimal proportions. Furthermore, they obviously may not contain lesser known or unknown nutrients which are, however, provided by natural foods.

—*U.S. Army Nutritional Manual,* 1949

In my opinion it is a much wiser policy to consistently educate the public in the use of proper combinations of our common, natural, wholesome foods, so as to make diets which are complete, rather than rely upon synthetic vitamins.

—Dr. Elmer V. McCollum, Professor of Biochemistry, Johns Hopkins University, *Science Newsletter,* 31 July 1943

[No] enrichment program can lessen the real importance of a properly selected diet of natural foods of high nutritional quality. In fact there is danger that the addition of small amounts of a few minerals and vitamins may create a false sense of security by obscuring the importance of the many other equally significant dietary factors, known and unknown. If the practice of enriching the finished product with a few nutrients should result in a trend to further "refinements" in manufacturing or in less attention to the conservation of nutrients in processing, the result might be disastrous. More can be accomplished nutritionally by chemists in conserving nutrition values in the processing and marketing of our basic food supply than in the enrichment of the finished products, or in making available vitamin and mineral supplements, important as the latter may be in special situations.

—Dr. William A. Albrecht, Chairman of the Department of Soils, University of Missouri, "Soil Fertility and the Human Species," *Chemical and Engineering News,* 25 February 1943

〉〉〉〉〉〉〉〉〉

Naturally Derived versus Synthetic Vitamins— Are They Identical?

Spokesmen for the FDA pontificate that there is absolutely no difference between synthetic and natural vitamins or other food supplements. In fact, the agency has proposed a regulation that would forbid a manufacturer from making any label or advertising claim that a vitamin product made from natural ingredients is better than a similar product made synthetically.

Repeatedly, FDA officials as well as members of the drug trade and some professionals keep assuring consumers that vitamins and other nutritional supplements are identical, regardless of whether they are derived from natural or synthetic sources. Apparently these individuals have missed or have chosen to ignore some important findings in the literature of their professions and simply spend their time quoting each other.

If "identical" refers only to the chemical composition, in a limited sense, the statement may be true. A molecule of crystalline ascorbic acid synthesized in a laboratory may be chemically identical in structure to a molecule of ascorbic acid derived from a natural source such as citrus fruit. However, the comparison should not be restricted to a simple determination of the chemical composition. And as the trend continues toward a diet comprised of more and more highly fabricated, factory-modified, highly processed foods, infant feeding formulas, and fortified breakfast cereals, as well as agricultural practices based on human "improvements" over nature, it is especially important that the subject be examined carefully.

Distinct and important differences do exist between natural and synthetic vitamins. Although molecule for molecule, synthetic and natural vitamins may appear chemically identical, distinct differences can be demonstrated with polarized light. The plane of a beam of polarized light is rotated by all natural substances either to the left (l-form) or to the right (d-form). Substances not occurring in nature may not cause rotation of the plane of polarization. These substances are said to be optically inactive (dl-form). What is the significance of these different forms of the same basic substance?

Natural versus Synthetic: More Critical Differences

Natural food protein, as well as protein in the human body, contains only l-forms of amino acids. The body possesses solely l-enzymes. Natural l-amino acids are absorbed immediately in the digestive system by the l-enzymes. Absorption is delayed when d- or dl-forms of amino acids are ingested. In fact, it is questionable whether the body can utilize dl-amino acids efficiently. Nevertheless, *dl-forms of amino acids are being used to fortify processed food, and are being used increasingly to boost protein of low biological value as well as to supplement animal feed.*

In July 1973 the FDA dropped dl-amino acids (synthetic) from the GRAS list, admitting that uncontrolled uses of synthetic amino acids to fortify foods "may result in risk to the public health from excessive intake of free amino acids." Animal studies have shown that an excessive intake of amino acids and amino acid imbalance can retard growth, lead to degeneration of certain organs, and lead to early death in animals.

Contrasting views were expressed on the use of the natural l-form versus the commercially available mixtures of dl-forms of amino acids. Because of a substantial lack of information about the safety and biological effectiveness of the dl-forms, the FDA maintained a cautious attitude and concluded that the acceptable forms of amino acids should be restricted to the l-form (except for dl-methionine and glycine, judged acceptable, but not for use in infant foods). The agency prohibited the use of acetate and sulfate forms of amino acids at present, since they "have no history of safe use." The FDA concluded that it is in the best interests of the consumer to permit and encourage "rational fortification" of food with amino acids by limiting supplementation to a safe level.

At least one scientist went on record opposing the entire concept of amino acid fortification and indicated that improvement of protein quality by fortification with free amino acids "may create an imbalanced diet and undesirable effects on human physiology."

Pantothenate, another nutrient, demonstrates distinct differences with various forms. The complete chemical structure of calcium pantothenate was determined when a method was devised to extract it from liver. All of the configurations of molecules were found to be d-forms. Pantothenates, having only slightly different atomic arrangements, were found to be ineffective. The l-form, for example, was termed useless.

Another critical difference between synthetic and natural vitamins is that a synthetic vitamin contains one pure substance or at most a few pure ones. It lacks other constituents present in a natural one that may possibly be vital for its full effectiveness. For example, the synthetic form of ascorbic acid (Vitamin C) consists solely of the isolated pure crystalline substance. The natural form, derived from foods such as citrus fruits, rose hips, acerola cherries, or black currants, has accessory factors present, such as other vitamins, minerals, trace minerals, enzymes, coenzymes, and other important nutrients. These substances, in combination with ascorbic acid as well as each other, may exert marked effects. Such interactions may determine the degree to which the ascorbic acid is absorbed and utilized. For example, physicians who use megavitamin therapy (use of vitamins in massive doses for therapeutic purposes) discovered that by using vitamins derived from food sources they were able to prescribe lower doses than when they used synthetic ones. The effectiveness at lower doses may be due to the presence of the important accessories, sometimes called naturally associated synergistic factors.

Some physicians have reported that naturally derived vitamins are not likely to overstimulate the glandular system, whereas a single fraction, in synthetic form, may be excessive and out of balance with other fractions of the vitamin complex.

Since vitamins are well diluted by the plant or animal substance in which they are found, there is generally little danger of ingesting toxic amounts of naturally derived vitamins. The notable exception is the exceedingly high concentration of Vitamin A in polar bear liver. However, high doses of synthetic vitamins are potentially hazardous since they are concentrated.

One researcher discussed the problem associated with synthetic vitamins that may be closely related to, but not identical with, natural ones: "The close relations, although useful in many ways, pose some problems in that they may have only a fraction, whether large or small, of the biological activity of the natural product." Synthetic vitamins may perform some of the useful functions of their natural counterparts but be useless for others. Hence, if used at all they should be used with knowledge and skill. For example, synthetic ascorbic acid may be effective as an antioxidant, but since it lacks the accessory factors of bioflavonoids that are present in naturally derived ascorbic acid, synthetic ascorbic acid is incapable of promoting capillary health.

Another critical distinction between synthetic and natural vitamins is the manner in which they are handled in the body. Crude natural extracts of vitamins are released more slowly than concentrated synthetic ones, which allows time for them to undergo digestive processes before they are absorbed and utilized. Since they are less soluble before digestion than synthetic ones, they can be more effectively deposited in the tissues. Most vitamins appear in tissues as components of complex coenzymes. Thus, the rate of dissolution and increased level of riboflavin in the blood will closely match the rate of its removal from the blood for coenzyme formation. In laboratory tests, using United States Pharmacopeia stomach-intestinal juice, it was demonstrated that pure synthetic riboflavin (Vitamin B_2) went into solution ten times faster than the same amount of riboflavin from yeast or liver. The sudden higher serum value derived from a rapidly soluble substance such as synthetic riboflavin does not necessarily result in higher tissue absorption of the vitamin. It is possible that the rapidly soluble substance exceeds the kidney threshold and the riboflavin is merely excreted in the urine. Anyone who has taken a tablet of pure riboflavin can testify to this.

Vitamin E: Tocopherols with Active and Inactive Fractions

Commercially prepared natural Vitamin E concentrates, derived from edible vegetable oils or from the by-products of their refining, rotate the plane of polarization to the right and hence are d-form. D-alpha tocopherol (natural Vitamin E) has been found to be considerably more active than dl-alpha tocopherol (synthetic vitamin E). Different forms of Vitamin E show very different biological activity, even though they appear to be essentially the same ring system and have similar scavenging properties in relation to "free radicals" (short-lived nonionic highly reactive compounds that may interfere with health or shorten life). All synthetic tocopherols have in one way or another been shown to produce different physiological effects. Some of the synthetic ones differ from d-alpha tocopherol in how they transfer across cell walls and accumulate in desirable concentrations where they are needed. To date there is no completely nontoxic synthetic substitute for d-alpha tocopherol.

Studies of the relative potencies of Vitamin E supplements have demonstrated that natural forms may be 20 to 36 percent more active than synthetic ones. A two-year feeding study of rats showed that naturally derived Vitamin E was far more potent than a synthetic one.

Naturally Derived Vitamin C versus Crystalline Ascorbic Acid

Medical literature contains numerous reports suggesting that naturally derived vitamins are often more effective than their synthetic counterparts. Frequently, ascorbic acid from natural sources has been demonstrated to be more biologically active than synthetically manufactured crystalline ascorbic acid. The naturally derived sources are better stored in tissues and organs, are better utilized by the body, and have more therapeutic value.

When synthetic ascorbic acid is used, it is more effective if it is taken along with foods in which ascorbic acid occurs naturally. This was recognized as early as 1937. Three out of twenty-nine humans suffering from scurvy failed to respond to synthetic ascorbic acid treatment. Recovery was achieved only after it was supplemented with fresh lemon juice. Prisoner experiences during World War II confirmed these findings. Studies made with war camp prisoners in Germany showed that scurvy conditions failed to respond to synthetic ascorbic acid therapy alone but cleared up with the addition of quantities of green onions. Similarly, synthetic ascorbic acid failed to cure gum inflammations among Norwegian troops until fresh, green, leafy vegetables were eaten. Later, experiments elsewhere also demonstrated that ascorbic acid, consumed in its natural medium of citrus juices, leafy greens, rose hips, green walnuts, and green onion tops, is better utilized in cases such as scurvy than is the synthetic form.

Vitamin A: A Complex

In nature, Vitamin A is a complex with a variety of forms. Preformed Vitamin A is found in animal food. Provitamin A (a precursor, called carotene), found in plant food, is converted within the body of the animal that consumes plant food containing this vitamin. Each form is distinct in chemical structure and scientific classification. Retinol, retinal, retinoic acid, and the retinyl esters comprise the preformed vitamins in Vitamin A_1, whereas dehydroretinal and dehydroretinyl esters are in Vitamin A_2. When retinol, a single chemical structure, is synthesized and labeled as Vitamin A, it is not the equivalent of the whole Vitamin A complex as found in fish liver, shark, or carotene oils.

Vitamin B: A Complex

A considerable body of literature exists demonstrating the superiority of using natural sources of Vitamin B complex over using single, synthetic fractions of the vitamin. As early as 1945, one researcher warned:

> Progress in biological research and animal experimentation with isolated B fractions has temporarily outrun our clinical experience and sound judgment. Those of us who have gained experience in treating

deficiency diseases with yeast and wheat germ before the synthesis of the various members of the vitamin B complex have certainly been disappointed with single or combined synthetic fractions. Examinations in man under controlled conditions have revealed that many polyneuropathies cannot be cured by any combination of the available pure vitamin products but only with yeast or wheat germ.

When this statement was made, a study had already been conducted comparing the relative values of supplements in the treatment of malnourished infants. A mixture of yeast, potato, banana, and dextrose maltose (called vegamine) was found to be vastly superior to the same mixture in which the yeast had been replaced by synthetic vitamins.

Similar results were confirmed in animal studies: "When all known synthetic members of the B complex were given, the animals failed to grow, their coats became depigmented, the fur quality deteriorated, and the animals died. These changes could have been prevented by the inclusion of such natural or complete source of the vitamin B complex as yeast or liver."

Another study concluded, "The results indicate the presence in yeast of some factors (or factor) which, together with vitamin B_{12} but not alone, bring about greater utilization of carotene and larger storage of vitamin A in the liver and kidneys of rats than occurs when vitamin B_{12} is fed with a synthetic vitamin-mix diet."

Pyridoxine, another Vitamin B fraction (B_6), was noted as "not metabolically active as a coenzyme unless in the form of pyridoxal phosphate or pyridozamine phosphate." These two forms are contained in natural foods, whereas commercially synthesized B_6 is pyridoxine hydrochloride.

Similar distinctions between natural vitamins in food, and synthetic ones fabricated in laboratories, apply to other fractions of the B complex. The forms that contain active coenzymes are found in foods.

Vitamin D: A Complex

Certain differences between natural and synthetic forms of vitamins are clearly defined with the Vitamin D complex. Although

Vitamin D activity may be activated by more than twenty different plant sterols exposed to ultraviolet light or by chemical reaction, the two forms most frequently used are Vitamin D_3 (natural) and Vitamin D_2 (synthetic). The structural forms of Vitamins D_3 and D_2 are quite different: D_2 has an unsaturated side chain, and the synthetic form is described as "rather more toxic than the naturally occurring animal vitamin."

Vitamin D_3 (cholecalciferol) is produced in human beings in the layers under the skin's surface. It is absorbed by the body after exposure to sunshine. It is also found in activated animal sterols and in certain fish oils such as cod liver or halibut. It is not yet well understood how deep-sea fish, not exposed to ultraviolet light, can produce their rich supply of Vitamin D_3. Prior to 1970, it was believed that Vitamin D_3 exerts its physiological functions directly, as do other vitamins. More recently, Vitamin D_3 has been regarded as a steroid hormone, metabolized to other chemical entities that are the active forms of the vitamin. The metabolites, produced in one particular organ and transported through the bloodstream, go to various sites to function.

Vitamin D_2 (calciferol) is prepared by irradiating ergosterol, a vegetable sterol, to develop Vitamin D activity. This synthetic vitamin is recognized by the familiar words "irradiated ergosterol" appearing on the labels of many food containers.

Vitamin D_2 was successfully synthesized in the 1920s, when rickets in children was prevalent. Fish oils were recognized for their antirachitic property, but food processors began fortifying many foods with Vitamin D_2 with wild abandon. Dairies fortified regular, evaporated, and powdered milk. Other food processors began to fortify baby foods, breakfast cereals, bread, macaroni, margarine, ice cream, prepared vegetables for infants, cocoa, and even hot dogs.

Since all forms of Vitamin D can be stored for long periods in the human body, none should be consumed in excessive quantity. As early as 1930, one consumer organization warned of the dangers of excessive intake. Infants were especially likely to develop hypercalcemia, an excess of calcium in the blood. Other warning signs had been gleaned from animal feeding experiments. In 1932, chicks fed synthetic Vitamin D_2 died, though chicks fed natural Vitamin D_3 thrived.

The toxic nature of Vitamin D_2 was well recognized in Great Britain, where infant hypercalcemia had been accidentally induced in a great number of infants due to the practice of fortifying infant feeding formulas, powdered milk, and cereals intended for infants. In 1957, after official action had been taken so that fortification of foods with Vitamin D_2 was substantially reduced, the incidence of infant hypercalcemia was dramatically lower.

In the United States, repeated warnings had been sounded in medical journals, but the FDA failed to take action for several decades. In the 1960s the agency requested a review, long overdue, by a joint committee of the American Medical Association and the American Academy of Pediatrics. On their joint recommendation, the FDA ordered a reduction of Vitamin D fortification in foodstuffs. But in July 1968, inexplicably, the FDA withdrew its proposal to remove Vitamin D from the GRAS list, though the vitamin should have been removed. By December 1972, the FDA proposed to limit the strength of Vitamin D products available for over-the-counter sale, recognizing that excessive amounts can be toxic.

Both D_2 and D_3 forms are available. Despite the greater hazard of the synthetic form, it remains the common choice in food fortification programs (e.g., milk fortified with Vitamin D). Vitamin D_2 costs less than the natural form, but due to the small amounts required, the cost is negligible. In some instances, dairies use the Vitamin D_3 form. Consumer pressures might influence food processors. Interestingly, the Vitamin D_3 form is used in some stock diets for such laboratory animals as monkeys and rats, but seldom in commercial foods intended for human consumption.

What Consumers Need to Know before Buying Vitamin Products

A vast and confusing array of vitamin products is available in the marketplace. In order to make intelligent choices, consumers need to read labels carefully and understand them.

The terms *natural* and *synthetic* are inadequate for classifying some commercially prepared vitamins. For example, a product, "Vitamin C from Rose Hips," may be labeled "natural." The product may be natural insofar as it contains rose hips that have been harvested, ground, and produced into powder. However, the product may not necessarily consist of 100 percent rose hips; it may also contain Vitamin C synthesized by fermenting glucose (a natural sugar obtained from corn or potato starch) with many steps of organic synthesis, so that the *process* should not properly be termed "natural." A Vitamin C product labeled "Vitamin C with Rose Hips" or "Vitamin C with Acerola Cherries," and so on does not consist of 100 percent rose hips, acerola cherries, or whatever, though such products may still be good sources of Vitamin C. The presence of some natural source will act synergistically to enhance the total effectiveness of the product. Many vitamin products labeled "natural" contain synthetic vitamins added to the natural base product. The Vitamin C in Rose Hip Vitamin C Tablets may be adjusted from an original 2 percent to 50 percent by adding synthetic ascorbic acid. If this were not done, the tablets would be prohibitively expensive and too large in size for the desired potency.

"From natural sources" should refer to the immediate origin of a product. To use this phrase as in "Vitamin A, from natural sources: lemongrass" is misleading, since thirteen synthetic steps are required to process the lemongrass, even though it is a natural source.

In order to make an intelligent choice in purchasing Vitamin E products, it is necessary to understand a number of terms (see page 144). The sole word *alpha* on the label, without a more specific designation, does not identify the source of the vitamin E. This is not in the interest of the consumer. If the label states, "d-alpha tocopherol (naturally derived)" or "dl-alpha tocopherol (synthetic form)," the vitamin is in its liquid, alcohol form, its most potent form, even though it is somewhat unstable in contact with air. *Acetate* means that the vitamin is in its liquid acetate form. This is a slightly less potent form, but is a more stable product. Acetate is an organic ester added so that the product resists oxidation and spoilage. *Succinate* means that the vitamin is in its solid succinate form, which is usually powdered for use in capsules and tablets. Succinate is also an organic ester added so that the product resists oxidation and spoilage. Tocopherol is spelled *tocopheryl* when the ester name follows the word, as in "alpha tocopheryl acetate" or "alpha tocopheryl succinate." *Mixed tocopherols* or *Vitamin E Complex* contain, in addition to alpha, three other natural tocopherols—beta, gamma, and delta—in varying proportions. These natural tocopherols are derived from vegetable oils such as soybean, corn, and cotton. In such mixed tocopherols only d-alpha is officially assigned Vitamin E values, since alpha is the one known to have biological activity. The other tocopherols are believed to have some biological activity in the body, but they are not yet well understood. Gamma tocopherol, plentiful in corn, has only about 10 percent of the biological activity of alpha. But gamma tocopherol has been shown effective in preventing peroxidation (which destroys oxygen processes) and it lowers the rate of red blood cell destruction. This feature has been noted in red blood count hemolysis tests.

Vitamin E, assayed solely in terms of alpha tocopherol, makes use of International Units (I.U.), meaning that if an oily-type Vitamin E 400 I.U. capsule is compared with a mixed Vitamin E 400 I.U. tablet, the oily-type capsule contains solely that and nothing more. The mixed tablet contains perhaps 20 to 30 percent free tocopherols that do not show up in the alpha assay test.

It is important to compare the relative biological activity of different Vitamin E products, as well as the number of International Units, and price. The following shows the activity of one milligram of Vitamin E in various forms:

d-alpha tocopherol	1.49–1.50 I.U.
dl-alpha tocopherol	1.10 I.U.
d-alpha tocopheryl acetate	1.36–1.40 I.U.
dl-alpha tocopheryl acetate	1.00 I.U.
d-alpha tocopheryl succinate	1.21 I.U.
dl-alpha tocopheryl succinate	0.89 I.U.
l-alpha tocopherol	0.50 I.U.
beta tocopherol	0.10 I.U.
gamma tocopherol	0.10 I.U.
delta tocopherol	very low

Natural versus Synthetic Vitamins

Some products contain blends of naturally derived and synthetic Vitamin E. Labels for such products should state the relative percentages of the d- and dl-forms present in the products. In reading vitamin product labels, the following will be a helpful guide:

VITAMIN	NATURALLY DERIVED SOURCE	SYNTHETICALLY DERIVED
A complex	Fish liver, shark, or carotene oils	Vitamin A palmitate, retinol
B complex	Yeast, liver, rice bran, or soybean; B_{12} (cobalamin) is from natural sources; to date it has not been synthesized	No complex; single fractions only, such as thiamine hydrochloride, riboflavin (other than yeast), niacinamide, pyridoxine hydrochloride, d-biotin, pteroylglutamic acid, choline bitartrate, etc.
C complex	Acerola cherries, citrus, currants, rose hips; the words *Vitamin C* should appear *before* the listed food sources	Ascorbic acid
D complex	Fish liver oils, irradiation of yeast or vegetable oils	Ipricated ergosterol
E complex	d-alpha tocopherol; d-alpha tocopheryl acetate; d-alpha tocopheryl succinate; mixed tocopherols; all naturally derived from vegetable oils	dl-alpha tocopherol, etc.

14. Mother's Breastmilk versus Infant Feeding Formulas

Pediatricians have successfully "readjusted" the milk of the cow to equal that of [wo]man. . . . Despite these efforts, genuine reconstruction of human milk from the bovine has not been wholly accomplished. . . . The apparent sturdiness of the bottlefed American citizen may, after all, be attributed to pediatric skill and Tender Loving Care, rather than the excellence of commercial "formulas" and breastmilk substitutes.

—George L. Fite, M.D., Senior Editor,
Journal of the American Medical Association, 8 April 1974

Infant formulas based on cow's milk or soy protein are adequate to support growth. As good as it may be, however, no commercial milk can rival a mother's milk. . . . Medical authorities agree that breastfeeding is the best way to nourish young babies.

—Charlotte Morin, *Canadian Consumer,* August 1975

It has been rightly said that a scientist who invented an infant food with the nutritional and antiinfective cost-effectiveness of human milk, which also solved the production and distribution problem in such an efficient way,

146

Mother's Breastmilk versus Infant Feeding Formulas

would be considered as a candidate for a double Nobel Prize—in medicine and in economics.

—Dr. Derrick B. Jelliffe, Professor of Pediatrics and Public Health, University of California, *statement,* "Maternal, Fetal and Infant Nutrition." Congressional hearings before a Select Committee on Nutrition and Human Needs, June 1973

Not only is breastmilk unique and impossible to imitate—despite manufacturers' claims—but the cost of cows' milk preparations remains beyond the means of the average family in the developing world.

—Dr. Derrick B. Jelliffe, World Health Organization Report, in *The New York Times,* 8 April 1973

U.S. Patent 3,896,240. Preparation of Simulated Human Milk: A process for the production of a composition that simulates human milk and which is especially suitable for the feeding of infants which comprises (a) treating whey with an anion-exchange resin charged with chloride ions to remove substantially all anions of weak acids from the said whey, (b) adding to the said treated whey a protein produced by pouring a preselected quantity of skim milk into a preselected quantity of an aqueous solution of hydrochloric acid.

—*DRINC* (Dairy Research, Inc.), September 1975

For their part, formula manufacturers have taken the position that the best food for an infant is the milk from his healthy mother. "Likewise, with proper care," says a nutritional research director for [an infant formula processor], "most infants can be properly fed with infant formulas." The trend toward breastfeeding, he notes, "has not dampened the company's sales."

—*Medical World News,* 13 September 1974

The superiority of breast over artificial feeding during the early months of infancy has long been recognized by the medical profession.

—Robert M. Woodbury, Ph.D., Director of Statistical Research, Children's Bureau, United States Department of Labor, "The Relation Between Breast and Artificial Feeding and Infant Mortality," *American Journal of Hygiene,* 1922

New facts about mother's milk keep turning up that suggest Nature, in instance after instance, is best left to her own devices.

—*Medical World News*, 16 June 1975

Cow's milk is no substitute for human milk. Cow's milk is splendid stuff—for little cows.

—Sir Ashley Montagu, quoted by Max Levin, M.D., *Current Medical Dialog*
(Baltimore: Williams and Wilkins, undated)

A pair of substantial mammary glands have the advantage over the two hemispheres of the most learned professor's brain, in the art of compounding a nutritious fluid for infants.

—Dr. Oliver Wendell Holmes, quoted in *The Womanly Art of Breastfeeding*
(Franklin Park, Illinois: La Leche League International, 1963)

〉〉〉〉〉〉〉〉

The Introduction of Feeding Formulas and the Decline of Breastfeeding

Perhaps the most profound change in child nutrition in the last half century has been the altered patterns of infant and baby feeding. Infant feeding formulas have largely replaced breastfeeding, and commercially prepared solids have been substituted for fresh home-prepared food for the growing baby. In accepting the use of such convenience foods, many mothers have lost touch with the basic nutritional needs of their infants and young babies.

According to Dr. Derrick B. Jelliffe, the general decline in breastfeeding in developed countries had its philosophical beginnings in the mid-nineteenth century, with a technological revolution in medicine. It became more and more common to consider anything mathematical or man-made as an automatic improvement over anything natural. Pediatricians came to accept the notion that a man-made feeding formula could be an adequate substitute for breastmilk. Hospitals developed a practice of separating mothers from their babies, a "mechanistic" approach seemingly designed "for quasi-military regimentation of patients and for the convenience of the medical staff."

At first, bottlefeeding was limited to upper-income families. They considered it scientific, modern, and fashionable. Women were impressed by the cleanliness of sterilized bottles. They believed the myth that breastfeeding caused breasts to sag and viewed breastfeeding as an inconvenience. Later, an increasing number of working-class women, laboring outside the home, accepted the use of infant feeding formulas as an economic necessity. In time, people of all social levels came to consider the traditional and beneficial practice of breastfeeding an exceptional, unnecessary practice that was old-fashioned and quaint.

Prior to the introduction of infant feeding formulas, 90 percent of all mothers breastfed. By 1946 only 38 percent of women leaving maternity hospitals nursed their infants; by 1956, only 21 percent; and by 1966 only 19 percent.

Although interest in breastfeeding in the United States has recently been revived, national surveys in 1971 showed that the interest is far from universal. Among college-educated women, 32 percent breastfed, compared to 8 percent among those with grade-school educations. Furthermore, after eight weeks, almost half the mothers discontinued nursing.

Infant formula manufacturers have vigorously promoted their products with health care professionals as well as prospective and new mothers, by distributing booklets, by putting up posters, by showing films and slides in hospitals, and by giving free samples of products to physicians and nurses so that they can pass them on to prospective buyers. In the past, companies tended to push their products as complete replacements for breastfeeding. But pressures exerted by nutritionists and such international organizations as UNICEF, WHO, and the UN's Protein-Calorie Advisory Group persuaded these companies to suggest using formulas as a supplement to breastmilk. Much of their promotional material for new mothers is now primarily instructional and provides useful child-care information.

Nevertheless, critics claim that the ubiquitous presence of company materials constitutes what Dr. Jelliffe calls "manipulation by assistance and endorsement by association." The end result is subtle but powerful discouragement of breastfeeding.

Although many individual physicians have recommended breast-feeding, the medical profession as a whole has stood on the sidelines throughout the period of decline in breastfeeding.

Differences between Breastmilk and Feeding Formulas

Although claims are made that babies thrive regardless of whether they are given formulas, breastmilk, or a combination of the two, many distinct differences exist in the nutritional composition of cow's milk (or soy) and human milk. These differences should be understood by physicians, nurses, hospital administrators, and, above all, prospective parents.

Numerous differences in the basic composition of human and cow's milk continue to be discovered. Findings demonstrate the uniquely complex nature of human milk, not always appreciated in former times. More than a hundred known constituents are present in human milk in proportions and chemical compositions that are distinctly different from the proportions and equally complex compositions of milk from cows and other mammals. In the last decade, more than three hundred scientific papers have been published that deal with the biochemical properties of human milk. Yet the incompleteness of knowledge in this field is indicated by the discovery as recently as 1966 of six heretofore unidentified polysaccharides in breastmilk.

Jelliffe emphasizes that the specific nutritional and antiinfective properties of human milk are "quite beyond the possibility of replication in cow's milk formulas." The formulas are only "approximate imitations" of the principal known ingredients in human milk. The food industry's claims that their products are "humanized" or "just like mother's milk" are biochemically and nutritionally incorrect, said Jelliffe. "Indeed, in view of inherent differences in physio-chemical structure of even the main constituents and the complex mixture of other ingredients, this objective seems literally unobtainable."

What is best for the calf is cow's milk; for the piglet, sow's milk; for the foal, mare's milk, and so on, through the animal kingdom, according to a principle known as *species specific*. For the same reason, the human infant is best nourished with mother's milk, assuming that the milk is produced by a mother who is healthy and well nourished. What features make mother's milk superior?

PROTEINS

In human milk, whey accounts for about 40 percent and casein for the remainder of the total proteins. In cow's milk the ratio is reversed and distinctly different. Whey accounts for only about 20 percent; casein, about 80 percent. For the human infant, whey proteins are considered nutritionally superior to casein.

The differences in protein ratios are reflected in the curds formed in milk during its digestion. Curds formed by human milk are soft and flocculent and easily digested by the human infant. Unprocessed cow's milk forms large, firm curds difficult for the human infant to digest. For this reason, cow's milk needs to be modified through several processings, to reduce the curds' size so that cow's milk can be used in infant formulas.

Still, protein is better absorbed by the breastfed infant, since the larger stools formed by drinking cow's milk carry out some of the nutriment.

Certain nucleotides (substances indirectly helping protein synthesis for building the infant's body) found in human milk differ from those in cow's milk.

Human milk contains more cystine (an amino acid) than cow's milk, but less methionine (another amino acid). Since the human infant lacks the enzyme necessary to utilize methionine at this early age, the low amount of the nutrient provided in human milk is part of nature's plan.

FATS

Human fat is well absorbed in the intestinal tract of the human infant (about 92 percent), whereas butterfat from cow's milk is poorly absorbed. The two fats have different compositions, and their fatty acids are arranged differently. In human milk, palmitic acid is found in a different form from that in cow's milk, and in human milk this fatty acid does not interfere with the utilization of fat or calcium by the human infant. In cow's milk there is a greater quantity of fatty acids that are poorly absorbed by the human infant. In order to improve the fat absorption in infant feeding formulas, many manufacturers have replaced the butterfat with mixtures of vegetable oils that are absorbed nearly as well as human milk fat.

However, Dr. Samuel J. Fomon, a prominent pediatrician, and other pediatricians have questioned the wisdom of this practice. The vegetable oils are far lower in cholesterol than is human milk. In one study, eight ounces of milk in a commercial infant feeding formula contained only four milligrams of cholesterol, compared to levels in breastmilk ranging from twenty-six to fifty-two milligrams. The serum and tissue cholesterol levels of infants fed such formulas differ considerably from those of breastfed infants.

Fomon and others suggest that adverse effects may result from the substitution of vegetable oils for human milk fat. The "cholesterol challenge" during early infancy may be necessary in order to induce enzyme systems in the body to function properly for cholesterol control. It may be better for an infant to be fed moderate rather than low amounts of cholesterol so that later in life the individual can cope with foods that contain cholesterol.

Fomon also suggests that in the absence of enough cholesterol, the infant's ability to synthesize certain hormones, or the functioning of its nervous system, may be impaired.

Human milk is rich in polyunsaturated fats and their long-chain derivatives, as well as in other substances. All are vital for human brain development. Cow's milk contains only traces of these nutrients. Instead it is rich in saturated and monounsaturated fats.

Lipase (a fat-splitting enzyme), which helps the infant use free fatty acids for energy, is richly provided in human milk.

Linoleic acid (an essential fatty acid) is found in quantities four to five times greater in human than in cow's milk. A diet deficient in linoleic acid will slow down an infant's growth rate and produce dry, scaly, thickened skin.

VITAMINS

Cow's milk is low in some vitamins needed for human nutrition, including Vitamins A, D, and E.

Human milk has three times as much Vitamin A as cow's milk. Human milk supplies all of the B vitamins needed by the infant. Folic acid (a fraction of the B complex) is found only in very limited quantity in cow's milk and is better supplied in human milk. Human milk contains from two to three times as much ascorbic acid (Vitamin C) as cow's milk. What exists in cow's milk is destroyed by heat treatment in pasteurization.

Human milk contains about twice as much Vitamin D as cow's milk, though both may vary in composition according to the diet of the mother or the cow.

There is some controversy about Vitamin E. It is thought that most infants are born with a Vitamin E deficiency, since the vitamin does not pass from the pregnant mother to the developing embryo. But this deficiency is overcome within a few weeks after birth, especially if the infant is breastfed. However, Vitamin E is inadequate in processed cow's milk or formulas unless such products are supplemented.

The subject of Vitamin E is related to a practice already discussed: the substitution of vegetable oils for butterfat in infant feeding formulas. Human milk is three times richer than cow's milk in essential fatty acids that affect growth. In human milk the ratio between Vitamin E and polyunsaturated fatty acids (PUFA) is balanced ideally for the infant's requirements. In feeding formulas it is necessary to make adjustments, because of a disturbed ratio of Vitamin E to PUFA.

MINERALS

Breastmilk appears to give complete protection against calcium deficiency during the infant's first three weeks of life.

Lactose, a milk sugar present in larger amounts in human than in cow's milk, increases the retention of calcium. A proper balance between calcium and phosphorus is important. Cow's milk, which contains more phosphorus than does human milk, may overload an infant's kidneys and lead to hypocalcemic fits in a bottlefed baby.

153

TRACE MINERALS

In general, both the infant and the young rapidly growing child require relatively large amounts of all nutrients. This may also be true for trace minerals. At present, little is known about the concentration of trace minerals in human as compared to cow's milk. Some evidence suggests that human milk may be a better source than cow's milk for one trace mineral, chromium.

The zinc content of cow's milk is quite variable. It is by no means certain that cow's milk contains enough zinc to meet the requirements of a rapidly growing human infant.

The higher sodium content of cow's milk may result in diarrhea, or in hypertonic dehydration in a bottlefed infant exposed to excessive heat. One pediatrician reported that infants switched from breastmilk to formula feeding developed edema, due to the higher level of sodium. The condition responded well when the infants were placed on a low-sodium diet.

Although human milk contains little iron, it does contain nearly twice as much as raw cow's milk. Despite the low level of iron in human milk it is recognized that, in general, the breastfed infant does not become anemic, whereas the bottlefed infant—even one that receives and retains more iron because the formula has been supplemented—is still more likely to become anemic.

It is recognized that a cow's milk formula, supplemented with iron, may aggravate rather than diminish an iron-deficiency problem. Also, sensitivity to cow's milk protein can cause gastrointestinal bleeding, which in turn may lead to anemia.

When solids such as fruits and vegetables are introduced to very young infants, iron deficiency may be aggravated. Such foods are not good sources of iron. Cereals, also introduced at an early age, are poorer sources of iron than are meats. Various studies show that only 2 to 19 percent of the iron from vegetables can be absorbed, compared to 10 to 30 percent of the iron from animal proteins.

The breastfed baby rarely has a critically low hemoglobin level, even if solid foods are withheld for the first six months. La Leche League International reports that physicians do not find cases of anemia in infants who are completely breastfed for this length of time. The league also comments that none of the recent studies of iron-deficiency anemia in infants were with breastfed infants. The studies were made with infants fed various types of cow's milk formulas, some supplemented with iron, the controls without.

Iron-deficiency anemia is related to nutrients other than iron. Vitamin E is essential for the proper utilization of iron, and Vitamin E deficiency has been found to be a cause of hemolytic anemia in bottle-fed infants. When iron was added to the formula to correct or prevent anemia, the addition tended to lower further the Vitamin E serum level. Other modifications of the formula were necessary because of the low ratio of Vitamin E to PUFA. Thus, infants on feeding formulas may have low levels of Vitamin E, whereas those on human milk have normal serum tocopherol (Vitamin E) levels.

An adequate intake of ascorbic acid is also related to the prevention of iron-deficiency anemia. Higher levels of ascorbic acid in human milk satisfy the infant's need for this nutrient.

Adequate amounts of copper are relevant to the prevention of iron-deficiency anemia. Human milk contains about three times as much copper as cow's milk.

A copper deficiency in humans, found in infants fed cow's milk, was reported as early as 1931. However, the report was unconfirmed, so subsequently it was thought that humans were not at risk from copper deficiency. This belief is no longer tenable. Copper, closely associated with iron, is necessary for the formation of hemoglobin. When there is a copper deficiency, it may lead to iron-deficiency anemia. Copper also plays many other vital roles, since it is closely associated with enzyme systems. In the last decade, copper deficiency has been identified as a not uncommon complication of malnutrition and diarrhea in infants, especially when they are treated with a modified copper-poor cow's milk preparation.

More recently, copper deficiency was noted in rapidly growing premature infants who were being fed on copper-poor milk formulas.

Human milk has a critical ratio of zinc to copper, which is lower than the zinc-to-copper ratio found in most foods, including cow's milk. According to the research of Dr. Leslie M. Klevay, the lower ratio in human milk may bestow a health benefit on the infant. An individual breastfed as an infant may be less susceptible to atherosclerosis as an adult, due to the low zinc-to-copper ratio.

Some Nutritional Deficiencies in Feeding Formulas

An episode in the 1950s pointed up the fact that infant formulas might be nutritionally deficient. More than fifty infants between five weeks and four months of age developed convulsive seizures. All had been fed solely on a canned evaporated milk formula. Upon investigation, it was found that during the heat processing of the formula, half of the pyridoxine (Vitamin B_6) was destroyed. One person who had been fed this pyridoxine-deficient formula in infancy, at the age of twenty-two years brought suit against the manufacturer, claiming that the convulsions she had suffered from the infant formula resulted in permanent physical and mental disability. She won her case.

In the early 1960s extensive skin eruptions were reported in infants suffering from phenylketonuria (a genetic anomaly marked by an individual's inability to handle phenylalanine, a food nutrient). These infants had been fed formulas especially devised with low phenylalanine content. Nutritional deficiencies were suspected. Young rats fed these formulas failed to maintain normal growth and health even after extra phenylalanine had been added. Then the formulas were supplemented with choline, riboflavin, Vitamin E, and calcium pantothenate. With these additions to the formulas, the animals' growth rate was nearly normal and the skin eruptions cleared up in a few days.

In the mid-1960s, infants requiring special feeding formulas—such as formulas low in lactose, calcium, or sodium—developed various ailments, including rashes, extensive shiny eruptions on the buttocks, cracked lips, and fissures at the angle of the mouth and the outer canthi of the eyelids. Although attention had been given to the special needs of these infants, the overall nutritional inadequacies of their formulas were recognized only after the death of two infants. As a result of this episode, researchers recommended: (1) that before such formulas are placed on the market they should be assessed by testing their effects on the growth of young laboratory animals; (2) that the exact composition of synthetic foods such as infant formulas should, as far as is known, be readily available; and (3) that if such synthetic foods "be incomplete for adequate nutrition, this should be clearly stated, with precise instructions on the supplements required."

More recently, other nutritional shortcomings of infant formulas have been recognized for special needs such as those of premature infants. When they were fed formulas deficient in Vitamin E, they suffered from edema and hemolytic anemia (a premature destruction of red blood cells). Experiments showed that these health problems could be prevented by giving the bottlefed premature infants a supply of Vitamin E. The discovery prompted the FDA to require an adequate Vitamin E content in infant formulas.

An examination of infant feeding formulas intended for full-term babies led to a recommendation that Vitamin E should be added routinely to all infant formulas, since processing is very destructive to this vitamin. Lack of Vitamin E in these formulas can cause serious stress reactions, from the fat peroxides that develop in the absence of Vitamin E and are highly toxic to various enzyme systems in the body.

Infant formulas may lose nutrients during long periods of storage. Spot checks in groceries and drugstores revealed that many of these products had been on the shelf for more than eighteen months—the maximum length of time considered acceptable by manufacturers. Investigators reported that the protein had coagulated in the containers of some outdated formulas and that vitamins and other nutrients had deteriorated.

Some Relationships between Feeding Practices and Health

Breastfeeding the human infant immediately after birth, as it occurs naturally in the animal world, has recognized physical and emotional benefits for both infant and mother. Colostrum, a special fluid produced by lactating women during the first three to five days after childbirth, is rich in immunoglobulins, antibodies that protect the infant against pathogens that may enter the gastrointestinal tract. Colostrum also contains white cells in a concentration as great as in blood. Research as recent as 1975 indicates that these white cells offer excellent protection against infection, notably necrotizing enterocolitis, a widespread and often fatal infection of the intestine.

The intestinal flora of breastfed infants are dominated by *Lactobacillus bifidus*, whereas the bottlefed infant has less of this beneficial flora. The bifidus factor promotes growth and helps the infant develop its own infection-fighting bacteria. Bottlefed infants, who miss the natural protection afforded by breastmilk, are more prone to many disorders, including diaper rash, colic, diarrhea, colds, staphylococcus, allergies, and even polio.

Numerous studies demonstrate that breastfeeding reduces morbidity and mortality in infants. In one extensive study, in Liverpool, more than twice as many bottlefed babies suffered illnesses as breastfed ones, and nearly six times as many bottlefed babies died.

A recurrent middle-ear infection (otitis media) is more common in bottlefed than in breastfed infants. The infant, given the bottle in a supine position, needs to apply a tremendous amount of suction while reclining. The alternating changes in pressure and muscle activity propel the milk slime from the nasopharynx to the middle ear through the eustachian tube. Otitis media is far less frequently encountered in the breastfed infant, who is cradled in a more or less upright position.

A dental scientist has suggested that breastfeeding can help prevent dental decay, whereas bottlefeeding encourages it. Infant feeding formulas contain added sugars. Bottlefed babies are being given a sweet-tasting, high-carbohydrate diet from birth and encouraged to develop a taste for sweetness. They grow into toddlers who crave sweets because of their early feeding experience, and an undesirable lifetime eating habit may be established.

The "nursing bottle syndrome" results in rampant tooth decay in young children who have been encouraged to suck on bottles for hours at a time long past the usual weaning age. Generally their nursing bottles are given as a pacifier at nap or bedtime. The sugar (or sweetened water, milk, fruit juice, or even soft drinks offered) in the feeding formula remains in contact with the teeth over long periods. Once called milk mouth or applejuice mouth, the result of the nursing bottle syndrome is decayed and rotted teeth, especially the upper front ones.

Whereas breastfeeding may assist in appetite regulation, bottle-feeding may contribute to obesity. The composition of breastmilk changes over the course of the infant's feeding. At the beginning of the nursing period, breastmilk is thin and watery. Later it is thick and rich. Analysis of breastmilk has shown that the milk at the end of the feeding period contains up to five times as much fat and protein as at the beginning. This change in composition may serve at least two different purposes. First, it helps to quench the infant's thirst, whereas bottlefeeding, with the rich cow's milk formula, may create thirst. Mothers often misinterpret the thirst as hunger and supply the infant with additional formula feeding, which increases the thirst. Second, the change in breastmilk composition may act as an appetite regulator telling the infant when to stop eating. The mechanisms that regulate adult appetite are not well developed in the infant. It is unlikely that the infant intake is regulated by the amount of milk in the breast, for at any given time the nursing mother's breast contains much more milk than her infant can drink. Nor is it likely that the infant stops eating because it tires of sucking, since very little action is required for the milk to flow. Rather, the change in the taste and texture of the milk may act as the infant's appetite regulator.

There is growing evidence that bottlefed infants are more likely to become obese children, with a lifelong tendency to excess body fat. The bottlefed infant is highly susceptible to multicellular obesity, a condition in which cells develop that retain fat even when rigorous weight control is monitored.

Behaviorist psychologists have observed that in the shift from breastfeeding to bottlefeeding the mother often tries to force the infant to finish everything in the bottle, since she has been told by her physician to provide the infant with a certain number of ounces of formula. Such insistence may contribute to the development of juvenile obesity. Overfed infants have been dubbed "pâté de foie babies."

There are also distinct health benefits for nursing mothers. Breastfeeding has always been recognized as giving a mother a feeling of essentialness and a sense of accomplishment, and the bonding between mother and infant results in a close, comfortable physical relationship.

In addition to these obvious psychological benefits, there are physical ones. In one study, the overall incidence of thromboembolism among women who breastfed was only half that of those who bottlefed. Breastfeeding helps to prevent anemia in the mother by delaying the resumption of menstruation for several months.

Breastfeeding delays a new pregnancy and thereby affords a natural spacing of children. This concept, regarded by some as an "old wives' tale," is confirmed by scientific studies. *Complete* breastfeeding—in which no solids or supplements are given to the infant during the first four to six months of life—has a definite effect on the natural spacing of children, since it usually tends to postpone the resumption of ovulation and the menstrual cycle for seven to fifteen months. Although it is not true that a woman cannot become pregnant as long as she nurses, pregnancy is extremely rare before the first menstrual period if she is completely nursing her infant.

Breastfeeding is the principal means of preventing hemorrhaging after delivery of the infant. Stimulation of the breasts stops the flow of blood in the uterus.

Recent medical research confirms the traditional idea that the woman who breastfeeds is less likely to develop breast cancer.

According to women who breastfeed, they get back in shape sooner and feel fit faster. Mothers have breastfed infants through the centuries. Breastfeeding is described by the La Leche League as a "natural and unique system of supply and demand which best serves mother and baby." In contrast, infant formula feeding, with its many shortcomings, has been described by one pediatrician as "highly experimental and uncertain."

15. The FDA Favors and Encourages Imitation Food Manufacture

〈〈〈〈〈〈〈〈〈

Imitation: something that is made or produced as a copy; an artificial likeness; counterfeit.

—*Webster's Third New International Dictionary,* unabridged

The National Academy of Sciences, in a new policy statement, warns that new and formulated foods which replace foods that make significant nutrient contributions should provide nutritional value at least equal to those that are replaced. Examples are foods that resemble dairy products, fruit juices and meats; or that serve as a meal replacement or snack food.

—*Nutrition Notes,* Fresh Fruit and Vegetable Growers Association,
September 1973

The FTC [Federal Trade Commission] would ban any statement that natural foods are superior to "junk" foods; that freshly-squeezed orange juice is superior to some artificially flavored, chemical concoction; . . . that protein from meat, fish, nuts, milk, vegetables, etc., is superior to synthetic protein made from petroleum, coal tar; and other misnomers.

—*Health Food Retailing,* October 1975

An attorney for the FDA . . . made this statement: "Labeling food as an imitation has had a bad impact on selling because people think it's inferior." "People think imitation food is inferior only because it is," writes Ruth Desmond [President, Federation of Homemakers].

—*Let's Live,* June 1976

It appears that the current regulation regarding imitation foods is a determination by FDA that nutritional labeling will not serve as an adequate guideline for consumers.

—Arthur D. Koch, cofounder and chairman of LABEL, Inc. [Legal Action for
Buyers' Education and Labeling], January 1975

As world population pressure causes the price of food to go up, there will be [an] increasing tendency to substitute foods of plant origin for foods of animal origin. FDA has prepared to adjust to this trend, insisting on just two primary conditions. The first of these is that there be no loss of nutritive quality in the substitution of the new foods for traditional ones, and the second is that the new foods be honestly represented so that the consumer knows what he is buying. The first condition is a technological challenge to the processor but one that is not too difficult to meet. The second condition is a challenge to the marketing organization, because it must somehow tell the consumer how the new food is to be used, without at the same time misrepresenting it.

—Virgil O. Wodicka, *FDA Consumer,* October 1973

The goal of the FDA's approach, as outlined and supported by the court, is "to provide consumers sufficient information on labels of food products so that reasoned and informed shopping decisions could be made."

—*Snack Food,* June 1976

[In George Orwell's *1984*] he spoke of doublethink—the power to hold two contradictory and opposite beliefs at the same time and accept both of them. . . . A popular orange drink is described as "natural orange flavor." Its ingredients list, in small print, sounds like a parade of food additives and artificial flavor and color. Natural equals unnatural. Real equals artificial. An old-fashioned lemonade is as new-fangled as a dictionary of additives. Old-fashioned equals new-fangled. Just like Grandma used to make. Maybe so,

The FDA Favors and Encourages Imitation Food Manufacture

if Grandma had a Ph.D. in chemistry. . . . Language is distorted beyond belief. . . . Government not only stands by and watches all this doublethink, doublespeak, and doubletalk. It facilitates and encourages it.

—Herbert Denenberg, Philadelphia *Sunday Bulletin,* 30 May 1976

Fabricated foods are not necessarily the nutritional equivalent of the natural foods they are intended to replace. . . . It may be difficult to be well-nourished and particularly to satisfy our requirements for certain micronutrients (some of which may not have even been discovered yet)—as fabricated and highly processed foods form an increasing part of our diet. With our present state of knowledge and the foods which are currently available, if one's breakfast consists of imitation fruit drink, egg substitute, and white bread, and one's other meals are similarly constituted, nutritional inadequacy may be likely.

—Jean Weininger and George M. Briggs, Department of Nutritional Sciences, University of California, *Journal of Nutrition Education,* October–December 1974

》》》》》》》》

The Imitation Labeling Controversy

Traditionally, federal agencies had a clear concept that *imitation food* was the term used for any food product that failed to comply with specific requirements. Hence, the USDA required the word *imitation* to appear on the label of sausages that contained excess cereal or water. The FDA required that the word *imitation* appear on the label of jellies or jams that contained less fruit than required for a standard product or for cheese with more water or less fat than specified.

However, by the 1960s what was meant by *imitation* became fuzzy. The marketplace was being flooded with new, formulated foods simulating traditional ones. Food processors who fabricated these products balked at having the term *imitation* applied because it had a negative effect on sales.

163

At the 1969 White House Conference on Food, Nutrition, and Health, it was recommended that the *imitation* section of the Food, Drug, and Cosmetic Act "should not be interpreted so as to become a trade barrier which would present a serious obstacle to the development and marketing of modified products with improved nutritional content."

Heeding the suggestion, the FDA decided to ease its time-honored rule of *imitation* labeling. Stephen H. McNamara, an FDA lawyer, said that "regardless of a food's nutritional merit, labeling as an 'imitation' has a substantial adverse impact on marketing because of the term's connotations of inferiority." The agency proposed a new regulation to permit the marketing of simulated foods without the pejorative label of *imitation,* provided that the new food was nutritionally similar to the traditional one and would bear a distinctive name that accurately identified or described its basic nature, such as "breakfast links" or "breakfast strips" instead of pork sausage links or bacon strips, or "pancake syrup" instead of maple syrup.

An added labeling implication of the new ruling was that any food fortified to bring it up to a required nutritional level would have to carry full nutritional information on its label; an imitation product would need none. McNamara reported that the "FDA sees this new regulation as a 'carrot' to encourage that new substitute foods be formulated so as to be nutritionally equivalent to their traditional counterparts," whereas a food sold as an "imitation" need not be nutritionally equivalent.

The new ruling drew criticism from several quarters. Some food technologists argued that this ruling, along with a proposal to define plant protein products, would inhibit the development of inexpensive plant protein foods by any requirement that they contain the same quality of proteins as the food they imitate. Since plant proteins are deficient in one or more of the essential amino acids present in animal proteins, this requirement would necessitate the addition of costly amino acids to plant-derived foods.

Consumer opposition, spearheaded by the Federation of Homemakers, charged that the new regulations could dupe consumers "into purchasing new substitute analogs containing numerous additives, plus the presently permitted vitamins and [minerals] in the mistaken faith that such contrived food products can completely replace natural foods."

Ruth Desmond, representing the Federation of Homemakers, quoted opposition from respected nutritionists who were concerned that "later nutritional discoveries may reveal these new food analogs tragically lacking in some vital substance or substances required by the human body for maintaining good health. . . . This would be a fraud on misinformed consumers—robbing them both health-wise and economically." The Federation of Homemakers brought legal suit, challenging the FDA's new interpretation of "imitation" foods.

This legal action was reinforced by John S. Dyson, Commissioner of Agriculture for the state of New York, who declared that imitation food could never be nutritionally equivalent to natural food, and sought a permanent injunction to prohibit representing imitation food products as equal to natural products. "Chemical soups" was Dyson's disdainful characterization of substitute egg, sausage, and bacon; imitation maple syrup; fabricated potato chips, and others. He charged that such products generally cost New York State consumers 40 percent more than traditional foods and provide less nutrition.

New York State law requires that synthetic products have to be labeled clearly as imitations of natural foods, and it bars any statements implying that such products are the equal of natural products when they are not. Dyson reported that his department had attempted to persuade the companies in question to change their labels and advertising, "but they laughed at us."

The court upheld the FDA's imitation labeling regulation. The Federation of Homemakers filed a new brief, contending that the clear definition of imitation foods, existing in present Federal Food Standards, be retained. Once again, the appeals court upheld the FDA and declared, "The regulation provides a safety valve for specific cases arising later in which nutritional equivalency and descriptive labeling do not adequately protect consumers from substitutes which are inferior in other ways."

Instead of making consumer knowledge about purchased food products easier, the ruling has made it far more difficult. Through the relatively simple process of adding vitamins and minerals to food products, food processors can now do away with the opprobrious "imitation" label. In instances where the term cannot be avoided, some food processors have skillfully exploited it. A new imitation mayonnaise, for example, contains approximately half the fat of standard mayonnaise. The product was promoted with the slogan, "We have to call it imitation . . . but you don't."

"FDA has done everything possible to pave the way for synthetic foods," charged Dr. Michael Jacobson, director of the Center for Science in the Public Interest. The FDA's relaxation of regulations for imitation food labeling represents one aspect of its policy. Closely associated to it was the program to launch nutritional labeling.

Nutritional Labeling Favors Fabricated Foods

Within the last few years, the FDA initiated more than fifty actions to upgrade the nutritional information on food labels. The agency reversed some of its former policies and now requires that food labels list nutrients as percentages of Recommended Daily Allowances (RDA) rather than the far smaller Minimum Daily Requirements (MDR) formerly used to specify vitamins and minerals needed to maintain health. How useful is this nutritional labeling?

No one denies that consumers have needed more adequate food labeling information. Ironically, the labels on pet foods, regulated by the USDA, actually had contained more nutritional information than labels on jars of baby foods, regulated by the FDA. According to the latter, the newer nutritional labeling will help shoppers identify more readily what nutrients they are buying with their food dollars. The labels give serving sizes, stated in common household measures; servings per container; calorie, protein, carbohydrate, and fat contents; and the percentage of RDAs, if it provides at least 2 percent of the RDAs. The label may also list polyunsaturated and saturated fats, cholesterol, and sodium contents. The type of fat is of interest to many shoppers, and this concession represents a sharp reversal of the FDA's former policy. Nutritional labeling can help with comparison shopping, within a highly restricted framework. To comprehend the limitation it is necessary to know how the concept of nutritional labeling developed.

The idea had its first real impetus in 1969, when nutritional labeling received the endorsement of the White House Conference on Food, Nutrition, and Health. One large food processor viewed nutritional labeling as a good selling point for its products and promptly developed a program. Its competitors fumed. They viewed the program as a merchandising gimmick and an attempt to "turn the whole charade into a marketing scoop." But several large supermarket chains, viewing nutritional labeling as a good marketing opportunity, quickly climbed onto the bandwagon.

Nutritional labeling was favored by processors of fabricated foods, since the manufacture of such products compels processors to know their exact formulations. Producers of basic commodities such as produce, meat, and dairy products were less enchanted, recognizing that the proposed nutritional labeling placed them, as well as consumers, at a distinct disadvantage. Such labeling would prevent producers of basic commodities from telling the truth about the nutritional value of traditional foods.

The regulations failed to take into account the wide variability in composition of natural, completely unprocessed fruits and vegetables. In the proposed regulations, the FDA allowed nutritional variations up to only 20 percent. Yet scores of studies show that variations in produce may be far greater, caused by genetic differences in plants; such factors as sunlight, temperature, soil, and water; and agricultural practices. One of many studies, brought to the FDA's attention by the plaintiffs, showed a difference of some 400 percent in the iron content of turnip greens grown on similar soil in two adjacent fields, planted with the same batch of seeds, and tended with the same fertilization and farm management. Such variations, contended the plaintiffs, do not short-change consumers in nutrients in fresh produce. Ample evidence demonstrates that, on the average, fresh fruits and vegetables provide the amounts of nutrients that they are estimated to contain.

The United Fresh Fruit and Vegetable Association, joined by Sunkist Growers, filed suit against the FDA to block enforcement of nutritional labeling regulations, which they termed "arbitrary, capricious, discriminatory, and a disservice to consumers." The plaintiffs argued that the regulations, if enacted, would prevent them from truthfully advertising the nutritional values of fresh produce since the regulations were framed solely to fit manufactured foods.

The dairy industry faced similar problems. Many studies of milk composition, sometimes listing ten to thirty nutrients or more, are provided by a few cows in carefully selected university herds. They may differ markedly from those of the average herd. The milk composition, as it actually exists on the market, may vary from day to day or batch to batch.

The meat industry, too, faced a dilemma. "Controlling nutrient content is not a major problem for fabricated food, but for . . . meats, one must take what nature produces."

Dr. M. J. Babcock of Rutgers University, in opposition to the nutritional labeling proposal, charged that it would favor the sale of synthetic foods, misrepresent nutritional values, and promote excessive fortification of foods.

George M. Briggs, a professor of nutrition, discussed some of the problems that would develop with nutritional labeling. "Many people are going to be misled into thinking that "good nutrition" is synonymous with eating a mixture of foods whose nutrients on the labels add up to 100 percent of the U.S. RDAs. This in itself is far from being good nutrition."

Briggs pointed out that the problem would exist mainly for those who depend heavily on mixtures of highly processed or manufactured foods such as imitation drinks, meat substitutes, pastries, and desserts and certain incomplete food supplements. "Many of these foods can be fortified with nutrients up to 50 percent of the new U.S. RDAs without special label claims. Combinations of such foods can readily show 100 percent or more of all the mandatory nutrients on the label. People will be misled to believe that they are eating an adequate diet," added Briggs, "unless they are aware of good nutrition concepts."

Briggs cautioned, "No person can survive on a diet containing only the eight mandatory nutrients or even when all of the nineteen vitamins and minerals are present which make up the total mandatory and optimal nutrients in the U.S. RDAs. A number of essential nutrients are not on this list." He mentioned specifically potassium and sodium and trace elements. Briggs suggested that "it would be foolhardy to eat diets not containing any of these nutrients until their place in human nutrition is fully understood. . . . Since all nutrients will not necessarily be present in manufactured packaged foods (even though they may contain all the U.S. RDAs), to insure the intake of all essential nutrients the consumer should continue to eat natural or traditional sources of food. To do otherwise would be folly. Life cannot be maintained for more than a few weeks, or several months at the most, on a diet containing only the U.S. RDA nutrients."

16. The New Look in Animal Feed

<<<<<<<<<

The most important single factor in determining the flavor of any kind of meat is the way you feed the animal which provides it. You can feed for flavor or you can feed for profit, but you cannot compromise; the two are incompatible.

—Waverley Root, "Taste is Falling! Taste is Falling!,"
The New York Times Magazine, 16 February 1975

Non-protein Nitrogen Feed Product and Method for Producing the Same. U.S. Patent 3,940,493. The method of producing a palatable, nontoxic and substantially neutral pH food product for feeding ruminant animals comprising the steps of: mixing together a predetermined quantity of an edible, ungelatinized starch-bearing food material . . . a predetermined amount of at least one non-protein nitrogenous substance selected from the group consisting of urea, biuret, ethylene urea and ammonium carbamate, and a predetermined amount of bentonite.

—*DRINC* (Dairy Research, Inc.), April 1976

The Texas Agricultural Extension Service fed a group of white leghorn hens a commercial feed containing a small amount of portland cement and found that the eggs laid were bigger and had thicker, stronger shells.

—*Vector Consumer Newsletter,* 13 November 1975

The New Look in Animal Feed

Process for Producing a Livestock Feed. U.S. Patent 3,870,798. A process of producing livestock feed from a raw garbage mass, comprising the steps of first refining said garbage mass by removing foreign, non-food particles, comminuting the garbage mass, then batch processing the garbage mass.

—*DRINC* (Dairy Research, Inc.), May 1975

A concept that has long been helpful to me in understanding agriculture is one I call molecular rearrangement . . . fundamental components join together to form the building blocks for plants and animals. . . . A seemingly obvious derivation of this concept of molecular rearrangement is the use of animal manures to feed animals . . . because manure is really a partially-used feed. Animals just haven't extracted all the feed value from it, so why not run it through again?

—Gordon Conklin, "Molecular Rearrangement" (editorial),
American Agriculturist, March 1976

Much of the potential of a [hybridized] supergrass is lost when grazed. If the pasture is stocked with the large number of animals required to utilize the lush growth, much grass is trampled. Contamination with waste, which often causes rejection by the grazing animals, is another problem that robs supergrass of its potential. Scientists [suggested a solution]: keep animals off the pasture entirely and feed them harvest grass . . . dehydrate it, and process it into pellets. Research indicates that this management system will produce 50 percent more beef per acre than grazing. However, harvesting, hauling, dehydrating, pelleting and feeding are expensive operations.

—"Meat Research, An Agricultural Research Service Progress Report,"
Information Bulletin No. 375 (Washington, D.C.: USDA, January 1975)

Allied Chemical Corporation claims one of its customers is trying to make life easier for cows. It adds sodium bicarbonate to rough foliage in an experiment to aid digestion.

—*The Wall Street Journal,* 27 April 1972

[USDA] scientists . . . have envisioned a portable grazing bunk that moves through a pasture, its speed electronically controlled to coordinate with the speed at which animals graze the grass. After the portable grazing bunk moves on, the stubble of grass could grow undisturbed until the cattle returned in less than one month. A spray boom would apply chemicals to the grass before it was consumed, to make it more digestible and to increase the animals' appetites.

—"Meat Research, An Agricultural Research Service Progress Report," *Information Bulletin No. 375* (Washington, D.C.: USDA, January 1975)

[Our] food animals are being manipulated like a lifeless industrial resource. . . . Our meat animals have been placed on a diet composed for the most part of medicated feed high in carbohydrates. Before they are slaughtered, these obese, rapidly matured creatures seldom spend more than six months on the range and six months on farms, where they are kept on concentrated rations and gain about two pounds daily. Our dairy herds are handled like machines; our poultry flocks, like hothouse tomatoes. . . .

—Lewis Herber, *Our Synthetic Environment*
(New York: Alfred A. Knopf, 1962)

Fish and shellfish are being raised on coal-slurry waste water from power plants, prime ribs and chicken are being produced from feed composed partly of sewage, and crops are being grown in fields irrigated with sanitized water flushed from toilets.

—*Express* (Easton, Pennsylvania), 26 August 1976

≫≫≫≫≫≫≫

Over the years, traditional sources of forage crops and grains for livestock feed have been replaced, in part or wholly, by less costly ones. Recently, the search for substitutes has been intensified. The rapidly rising costs and increasing scarcities of raw materials that markedly affect human foods also influence animal feed. Many feed replacers now being given consideration are ones formerly underused or rejected because of their low nutritive values, lack of safety, or poor palatability. Substances now being considered include not only agricultural wastes but industrial by-products, municipal garbage, sewage sludge, and solids screened from animal manure. Experiments are done in pursuit of such goals as totally synthetic diets, all-roughage diets, and nonprotein diets. The trend

172

brings to mind the classic story of the stingy farmer who resented the cost of feeding his horse. Day by day, the farmer reduced the rations. Ultimately, the horse dropped dead, much to the farmer's surprise. How far will we go with feed extenders or substituting poor-quality feed before the day of reckoning comes?

Replacer Feed for Calves and Piglets

Calves and piglets both lack immunologic protection at birth but derive it in the colostrum from cow's or sow's milk if they are suckled directly after birth. The gastrointestinal tracts of these animals are permeable to protein for only the first day or two of life, and the immunoglobulins in the colostrum are absorbed only if they are given to the animals during that crucial time.

Calves that fail to obtain a fair share of colostrum within the first day of delivery often develop an infection, colibacillosis, and may die of gastroenteritis or septicemia. Piglets that lack colostrum after delivery are also liable to gastrointestinal infections. Yet both calves and piglets may be taken off cow's and sow's milk and given reconstituted dry skim milk replacers. Dairy farmers often prefer to dispose of calves in order to market the milk. Others, anxious to build up herds, prefer to buy young calves. Hog raisers want their sows to produce the maximum number of litters in a minimum of time. By feeding skim milk replacers to piglets instead of sow's milk, the sow is released for reinsemination as soon as the piglets have been delivered.

Studies show that calves raised on replacers do not gain weight as well as those raised on fresh whole cow's milk. Gastroenteritis often develops, with a significant mortality rate. The effective treatment is whole cow's milk. Colienteritis may also develop in calves brought up on dry skim or heat-treated milk. Heating denatures the immunoglobulins, which are the protective factors in the whey protein.

Piglets deprived of sow's colostrum also fared badly. Experimenters substituted intraperitoneal gammaglobulin, derived from pigs, for the sow's colostrum. Although the substitute raised the immunoglobulins in the serum to near-normal levels, it failed to provide protection against colibacillosis. Even daily injections of immunoglobulins were ineffective. The piglets were poorly protected even if colostrum was fed to them for the first twenty-four hours but was not followed by suckling. Piglets raised on a cow's milk replacer contracted colibacillosis and often died. Only colostrum feeding, followed by four to six weeks of suckling—as nature intended—provided complete protection against colibacillosis. When suckling time was reduced to less than that, piglets developed troublesome, though not fatal, colibacillosis.

For optimum growth and protection from gastrointestinal infections, piglets and calves, as well as human infants, apparently require local and systemic protection. Obviously, the milk replacers used for calves and piglets are inferior to fresh whole cow's and sow's milk, as are the feeding formulas intended for human infants, which replace breastmilk.

Nonprotein Nitrogen (NPN) as Feed Replacers

Protein is essential for beef cattle. Plant and animal proteins, used traditionally in cattle feed, contain nitrogen in the amino acid form. Such protein is well-utilized by cattle, but it is expensive.

Compounds such as urea, biuret, diammonium phosphate, and ammoniated polyphosphates contain large amounts of nitrogen, but they are not in the amino acid or protein form. These nonprotein nitrogen compounds (NPN) are favored by cattlemen as inexpensive replacers for more costly traditional sources of protein feed such as soybean, cottonseed, or linseed meals. It is well established that animals thrive on the traditional sources of protein feed. However, in many commercial operations, urea is now being used as a *total* nitrogen replacer. This practice may appear to be economical to cattlemen, but it is a shortsighted measure. Experimentally, it has been demonstrated that animals fed urea as the sole nitrogen source had inefficient weight gains for marketing. Also, if cattlemen attempt to speed up the raising and finishing of animals for market, toxic overdosing with too much urea or other NPN products is commonplace. If cattle are given too much NPN rapidly, they gorge themselves and cannot digest the large amounts. Toxic symptoms range from lack of coordination, muscle tremors, breathing problems, bloat, and stiffness of the front legs to death.

Attempts have been made with dairy cows to feed the same quantity or less protein than usual, but to get the animals to produce more milk. One method is to treat feed with formaldehyde. Another attempt was to gradually replace the normal feed of "dry" cows— cows not milking—with a semisynthetic diet consisting of compressed briquets of purified starch, cellulose, sucrose, urea, and ammonium salts; with a wet paste rich in cellulose; a small amount of corn oil and Vitamins A and D; and later, also Vitamin E. At the beginning of the experiment, the cows were allowed to chew on rye or wheat grass, but later cellulose strips impregnated with silicic acid were substituted. The cows were given hard rubber tubing to chew in order to help them secrete saliva. Adequate saliva is crucial for digestion, since high levels of urea in the diet may impair salivation.

Cement kiln dust is being investigated by USDA scientists for use in cattle feed to induce significant weight gains. Farmers have traditionally known the impossibility of making a silk purse from a sow's ear. But agricultural technologists keep trying.

Roughage Replacers in Feed

Traditionally, roughage was supplied to animals from field crops such as alfalfa. Among current roughage replacers are organic materials such as feathers and nonorganic ones such as polyethylene plastic pellets. However, wood products are the most popular roughage replacers. Unused cellulose from wood is the most abundant organic compound on earth.

Fibrous residues from commercial pulp and paper mills may be used as animal feed if the lignin and crystalline structure of the cellulose in raw wood pulp are reduced or eliminated.

175

In experiments, steers fed up to 50 percent unbleached southern pine kraft pulp in their rations for 58 days maintained weight gains and appeared healthy. Ewes on similar rations appeared healthy and maintained their productivity even during lambing periods. Treated wood has also been used in animal rations. Hemicellulose, made by high-pressure steam treatment of hard and soft woods, was fed to sheep at levels of 5 to 50 percent of their rations. The wood mixture was unsatisfactory at the highest levels, but palatable at 30 percent. The feed was used satisfactorily with rams for five years, with no ill effects found in ewes through two reproductive cycles.

Ground-up aspen (a western poplar) trees were added to cattle rations. Aspen-fed steers gained weight twice as fast as their corn-fed counterparts and required less feed. The aspen-fed animals produced steaks with less marbling than grain-fed animals and the steaks were reported to be equally tender.

Wood chips, commonly available in large quantities, are under consideration as roughage replacers. Studies are being done to find methods of converting the discarded chips into products that can be used by animals capable of digesting such substances.

Sawdust has been suggested as an economical, nonnutritive roughage material in a high-concentrate ration. Its lignin content, which is indigestible, needs to be lowered to a level comparable with good quality roughage. Experimentally, pelleted pine sawdust, at levels up to 10 percent of the total ration, was fed to calves for 90 days without any symptoms of toxicity. No detrimental effects were noted either on feedlot performance (appetite and weight gain) or on carcasses of cattle fed high-energy rations for 115 days, when sawdust replaced half of the alfalfa roughage in a 90 percent concentrate ration. There were no abortions or lowered intake when first-calf heifers were fed a ration of 25 percent raw pine sawdust during the last third of gestation.

Rumose, a manufactured word derived from *ruminant* and *cellulose,* is being considered as a roughage replacer for dairy cows and possibly for rabbits, sheep, and goats. Rumose consists of discarded corrugated cardboard boxes that are chopped, ground, or pelleted and mixed with animal feed. But the possibility of contamination by polychlorinated biphenyls (PCBs) in recycled cardboard deserves serious attention. PCBs may be hazardous to animal health and may contaminate the milk and meat products from animals fed such rations.

Similarly, the practice of using newsprint as animal roughage may be highly objectionable. Studies of newspaper, magazine, and computer paper use as animal roughage showed that these materials contained potentially hazardous contaminants such as toxic heavy metals (mercury, boron, lead, barium, and antimony), especially the comic sections of newspapers in which colored inks are used.

With the problems of various environmental contaminants now being found in wood wastes, the enormous research effort being expended in order to feed these nonnutritive substances to animals is open to question. Many wholesome agricultural waste products— including sugarcane bagasse, rice hulls, cottonseed hulls, oyster shells, and corn cobs—could be exploited for ruminants to a much greater degree. At present, quantities of these safe and nutritious materials are underutilized or wasted on or near farms and feedlots. One critic remarked that such agricultural wastes offer no problems as a feed when mixed with high-grade feeds. In addition, public land alongside highways produces grass that could feed countless numbers of animals. Instead, the grass is allowed to grow and produce weed seed or is mowed and left to rot.

Animal Feed from Sewage, Sludge, and Garbage

Sewage produces protein-rich algae. A sewage treatment has been developed using genetically altered algae that reproduce at high efficiency, are rich in unsaturated fats and vitamins, and contain 50 percent protein. In the treatment process the sewage water cascades through tiers of small tanks seeded with algae. Removed by a centrifuge, the algae can be dried, sterilized, and used as animal feed.

Sludge is being considered for use in animal feeds. The drawback is its toxicity, the result of the presence of heavy metals, phosphates, and fluorines. Treatment with sulfur dioxide has removed such contaminants. In experiments, egg laying by hens was reduced when the birds were fed treated sludge as 5 percent of the ration. Rats fed at the same level were able to reproduce normally, but at 10 percent their fertility, viability, and lactation were reduced.

Scientists hope to develop a range of microbes suitable for converting almost any waste organic material "at a rate typical of the processing industries," and they hope to utilize protein from municipal garbage waste in animal feed. The economic trade-offs in support of the new process are reported to be favorable, because of the steadily growing costs of garbage disposal.

Dried Poultry Waste (DPW) in Animal Feed

The idea of recycling animal manure in animal feed is not new. As early as 1917, a USDA farmers' bulletin listed among the hog's attributes its ability to make use of nutrients in manure that cattle had failed to digest. With changes in farm management practices and an increase in animal diseases, the practice of feeding manure to livestock fell from favor. Now, concerns about protein feed shortages and searches for alternative feeds previously wasted or underused, combined with the high cost of animal waste disposal and the pollution problems it creates, have all helped reactivate interest in recycling manure in animal feed. Processes have already been developed to treat animal wastes by dehydrating and pelleting, ensiling, and fermenting.

The stimulus for renewed interest in recycling animal wastes in animal feed was provided to a great extent by poultry raisers. They discovered that the recycling of dried poultry waste (DPW) in animal feed was more profitable for them than its traditional use as fertilizer. Considering the price of soybean oil meal, poultrymen estimated that the manure from one laying hen, when properly processed as cattle feed, was worth a minimum of fifty cents per bird above the costs of processing the manure. The push began.

The most attractive use of DPW is with ruminants such as cattle and sheep. Because of the difference in the digestive tracts of such animals, they can use the nonprotein nitrogen compounds that chickens are unable to utilize. As much as two thirds of the protein requirements of cattle were found replaceable by properly processed DPW. However, there is a great variation in DPW quality, with protein levels ranging as low as 12 percent and as high as 45 percent. DPW has also been considered as an ingredient in rations for laying hens, but not for broilers, growing turkeys, or hogs, for all those require high-energy feed.

How efficient is DPW? In USDA experiments, steers appeared to thrive when fed rations consisting of 20 percent poultry manure to 80 percent grain, from an early age when the steers weighed about 250 pounds to a finishing age of about 1,200 pounds. Preliminary results showed no differences in weight or general health from control animals fed the same volume of feed consisting of cornmeal supplemented with cottonseed meal. But when more than 30 percent DPW was used in the rations, performance and carcass quality were significantly adversely affected, and the animals found the rations unpalatable.

In another set of experiments, DPW substituted for 12 percent and 25 percent of corn in rations fed to laying hens. One group of hens had their own manure go through them fourteen times. Their production averaged slightly lower than that of a control group that received no manure.

Proponents of DPW use argue that it will cut livestock production costs, but such savings are not necessarily passed along to consumers. Proponents suggest that DPW use will conserve natural resources. But animal wastes may be far more valuable as high-quality fertilizers or as sources of energy if converted to methane gas than processed into a low-value, potentially hazardous animal feed.

Although the use of DPW is opposed by some on esthetic grounds, the real issue is safety. Are food products from animals fed DPW safe for human consumption? Bacteria, viruses, veterinary drugs, antibiotics, hormones, and heavy metals are frequent contaminants of poultry manure. What assurance is there that the contaminants will not be in the food products?

The use of DPW from laying hens gained approval in several

states. But because of unanswered questions concerning safety, both the FDA and USDA prohibited interstate shipments of food products from DPW-fed animals. However, by 1976 federal regulations were relaxed. The FDA approved a limited use of DPW for cattle not being finished for slaughter (brood stock, overwintering cattle, and calves not old enough to go into feedlots). It proposed that all cattle fed animal waste and intended as human food must have DPW feed withdrawn for a period of sixty days before slaughter to allow time for the animals to metabolize any excessive amount in their bodies. Opponents do not feel reassured by these measures. Past federal monitoring programs have failed to check thoroughly residues of diethylstilbestrol (DES), arsenic, or antibiotics in the meat supply. Consumers are inadequately protected. Is there any assurance that monitoring of DPW residues will be better?

Replacer Feed for Oyster, Lobster, and Trout

Replacer animal feed is not limited to farmlands. Replacers are also used in intensive aquaculture systems for shellfish and fish production that depend on heavy feeding. Treated organic waste such as domestic sewage is being considered for use in oyster production. Minerally enriched waste water is of a composition similar to that of commercial fertilizer and has some of the same properties. It can be used to grow algae under controlled conditions. The algae, having consumed all or most of the nutrients in the waste water, are then fed to oysters. Ideally, the result is faster-growing, plumper oysters as well as waste water that has been treated by filtration through the oysters.

Feeding is the most expensive aspect of raising lobsters in a controlled environment. The most satisfactory food used to date is live brine shrimp; the next most satisfactory, frozen brine shrimp. However, the food has become so costly that by 1975 one dollar's worth of brine shrimp was required to produce four dollars' worth of lobster. A number of artificial foods for lobsters have been developed, some costing as little as nineteen cents a pound. Although lobsters find them palatable, they grow much more slowly on the substitute

The New Look in Animal Feed

foods. Scientists are trying to discover what the brine shrimp contains that makes it so satisfactory, and they then hope to duplicate it less expensively.

Radical changes have occurred in fish nutrition and culture. From the inception of the artificial propagation of trout in 1853, they have been reared almost entirely on meat diets. The increasing competition for slaughterhouse products gradually priced these meat sources out of the economic range of the trout producer. Beginning about 1930, an increasingly large amount of dry meals (primarily scrap marine fish and liver meals) and cereal grain products were added to trout diets. The first all-dry feed diets, which eliminated the necessity of refrigerating meat, began to appear in the early 1950s, in conventional pellet form, supplemented with vitamins, antioxidants, growth stimulants, and other additives.

The feeds developed for trout were similar to those used for poultry. When the pellets were fed to baby chicks, the birds developed cancer. In the early 1960s, hatchery-raised rainbow trout, fed this pelleted feed, developed liver cancer in epidemic proportions. In some hatcheries, 100 percent of the trout were affected. Ultimately, the problem was identified as a contaminant in the feed. These incidents illustrated some of the possible hazards in substituting highly processed feed.

Synthetic Nutrients in Feed

The search for protein to fatten livestock stimulated several large petrochemical companies to develop processes for making Single Cell Protein (SCP) from microorganisms grown on molasses, waste starch, petroleum, paraffin, natural gas, oil, or any other easily available substrate, such as the waste sulfites from paper manufacture. SCP looked promising not only as animal feed but also as human food. The Amoco Food Company now supplies processors of food intended for human consumption with a torula yeast, grown on a substrate of ethyl alcohol, a by-product of the petroleum industry. This food yeast is advertised as a "natural food ingredient."

However, after two decades of experimentation and development

181

with SCP, the early enthusiasm about this product as a panacea has been somewhat dampened.

The sharp rise in fuel oil costs has discouraged the conversion of oil to feed and food. The program has also had a series of setbacks due to unforeseen health hazards.

In 1973, two SCP producers in Japan had to shelve plans when trace amounts of a cancer-causing substance, 3-4 benzpyrene, were found in the product. Presumably the contaminant was a residue of the oil that had not been filtered out. Also, trace amounts of lead, mercury, and arsenic were found in the product. Both 3-4 benzpyrene and the heavy metals increased in concentration in the final product. Chronic toxicity tests, performed by the Japanese Ministry of Welfare, demonstrated that SCP, fed to animals, resulted in decreased body weight, with congestion and enlargement of the kidney, liver, and spleen.

A more recent setback occurred in Italy, where the government denied final approval for the marketing of an SCP product. The fat of pigs raised on feed containing 30 percent of a British petroleum-based SCP product contained 71 parts of paraffin per million parts of fat. The diet of the pigs contained three times more SCP than permitted by Italian regulations.

The consideration of SCP products as human food as well as for animal feed led an editorial writer in *The Lancet* to caution that we do not know what effects these products will have on humans in terms of toxicity.

> One has to be sure that there is no contamination of the product by microbial toxins or heavy metals, solvents and carcinogenic hydrocarbons from the growth and extraction media. Furthermore, the high nucleic-acid content of these microorganisms means that one should not eat too much, and individuals liable to gout could be particularly at risk. Slight changes in conditions of growth or isolation of the organism could alter toxicity.

Synthetic amino acids have been proposed as a substitute for protein from natural foodstuffs in animal feed. However, amino acids in a free state, isolated from foods, can be toxic.

In 1965 the USDA proudly announced a milestone: the birth of calves from animals reared on a totally synthetic diet. An Angus cow, Number 248, had not had a mouthful of natural feed since it had been weaned in November 1962—believed to be the longest period a ruminant has lived on a totally synthetic ration. It had been fed a diet containing urea as the only source of dietary nitrogen, supplemented with cornstarch, corn sugar, wood pulp (for roughage and energy), corn oil, minerals, and vitamins.

Cow Number 247 was her identical twin sister, fed on a diet of natural feeds: ground corn, alfalfa hay, orchard-grass hay, linseed meal, cottonseed meal, bone meal, and vitamins and minerals.

Both cows, while heifers, were started on the experimental diets when they weighed about 290 pounds and were about six months old. Equal amounts of energy and nitrogen feeds were supplied daily and feed intake was limited to keep the heifers gaining approximately one pound daily. Although a little reluctant to eat the synthetic ration at first, Number 248 responded satisfactorily and accepted its unusual diet in about two to three weeks.

Number 248 gave birth to a fifty-one-pound heifer calf, which appeared normal at birth. During the first sixteen days of its life, the calf gained twenty-nine pounds, about equal to the weight gained by the heifer calf born to Number 247 on a diet of natural feeds.

Both calves were sired by the same bull and both were nursed by their respective dams.

On the sixteenth day of its life the calf born to Number 248 was found by the herdsman lying dead in its pen, less than half an hour after the herdsman had observed the calf appearing vigorous and healthy. An extensive postmortem examination failed to show the cause of death. Body organs and fluids appeared normal. No nutritional deficiencies were found.

The beef cow on the totally synthetic diet appeared to thrive. She was capable of reproducing. But long-range, unanticipated problems developed. With our present knowledge, the nature of the biological damage may have been too subtle to be detected or explained. The problem arose only in the second generation. The old and oft-quoted phrase "Nature knows best" is still valid.

Toward a Rational Feeding Program

Current livestock feeding practices are being presented, falsely, as practices that will release grains to satisfy global human needs. That pretense hides the real motive behind the substitution of poor quality or hazardous feed: lowering feed costs for livestock producers. In any discussion, this primary motive must not be lost. Do these replacer feeds actually help alleviate world human hunger?

Through the centuries, livestock was fed on grasses. Grain feeding came into practice only when grain was abundant and relatively inexpensive as a forage replacer for livestock. Grain feeding achieved faster weight gains for animals and thus higher profits. In recent years, with higher grain costs and larger demands for it as human food, this trend has been reversed. The rational solution is in part to return to traditional forage feeding.

Grain feeding can be discontinued. Ruminant animals such as cattle can be produced efficiently because they utilize, to a great extent, quality feed products such as forage that cannot be eaten directly by humans. Much land unsuitable for intensive use can be used for forage. Livestock provide a means of using grassland crops and converting forage into meat and milk. The protein quality and taste appeal in human diets improves substantially when animal products are included. Why should we covet high-quality animal protein consumption for ourselves and at the same time be content to offer low-quality vegetable protein consumption (represented by grains and legumes) for others? It is ironic that at the very time we have begun to perceive the values of high-quality nourishment for optimum health, our agricultural practices lead us further and further toward foods of lower quality. Governmental agricultural experiment stations, where much of the work is being conducted, have been responsible for this trend.

At a 1975 conference dealing with livestock management and global food production, there was little disagreement about the need and vast potential for increasing livestock production in Third World countries. Several participants suggested that Latin America and Africa could be capable of large increases in meat and milk production without significantly cutting into other food crops consumed by human beings. Much of the world's rangeland is poorly used but offers opportunities for major increases in meat production. Areas that receive too little rain to support crops or are too hilly to be farmed can still be grazed or browsed by ruminants. Such regions constitute 40 percent of the world's total land area. One expert at this particular conference suggested that the only alternative that arid lands have to produce food is through livestock. Although, admittedly, much of the world's rangeland is overgrazed, better management, including various water conservation methods, could double or triple the meat productivity of this land. Producing meat through better use of rangeland has an additional bonus. Energy requirements are low, since animals do their own food harvesting. Traditional forage feeding could serve as a means of making available more high-quality protein foods needed by humans everywhere.

17. Real Food—An Endangered Species

<<<<<<<<<

Scientific technology today appears at first glance to be merely an extension, even though a spectacular one, of what it started out to be in the early nineteenth century. In fact, it is different in nature. Until a few decades ago scientists and technologists were concerned with well-defined problems of obvious relevance to human welfare. . . . Unfortunately, modern man developed new technological forces before he knew how to use them wisely. All too often, science is now being used for technological applications that have nothing to do with human needs and aim only at creating new artificial wants. Even the most enthusiastic technocrat will acknowledge that many of the new wants artificially created are inimical to health and distort the aspirations of mankind. There is evidence furthermore that whole areas of technology are beginning to escape from human control.

—Dr. René Dubos, *So Human an Animal*
(New York: Charles Scribner's Sons, 1968)

No field within the entire scope of mechanization is so sensitive to mishandling as that of nutrition. Here mechanization encounters the human organism (whose laws of health and disease are still incompletely known). The step from the sound to the unsound is nowhere so short as in the matter of diet. . . . If man deviates too long from the constant of nature, his taste

Real Food—An Endangered Species

becomes slowly vitiated and his whole organism threatened. Unwittingly, he impairs judgment and instinct, without which balance is so easily lost.

—Siegfried Giedion, *Mechanization Takes Command: A Contribution to Anonymous History* (New York: Oxford University Press, 1948)

It seems evident that dietary advice should take into consideration the availability of food items, their composition and their subtle impact on the nutrition and the biochemistry of the individual cell. The food industry has a vital role to play in furnishing new food items which have been thoroughly tested for more than their freedom from toxic compounds. New food items should be considered in terms of their total nutritional impact as such food items will become increasingly more important to human welfare in a world of rapidly shrinking supplies of food items which have served as the main sources of proteins, vitamins and minerals for countless generations.

—Dr. Fred A. Kummerow, *Journal of Food Science,* January–February 1975

FDA does not set nutritional requirements for food and has no authority to require manufacturers to produce food that is nutritious.

—*FDA Consumer,* November 1975

The introduction of new technology without regard to *all* the possible effects can amount to setting a time bomb that will explode in the face of society anywhere from a month to a generation in the future.

—Elmer W. Engstrom, *American Scientist,* 1967

There is danger in combining the products of human ingenuity in the matter of food processing and food preparation with the capacity to develop liking for foods that are so defective in essential elements that when they are made a preponderant part of our diet, we may develop serious malnutrition.

—Dr. Anton J. Carlson, Professor of Nutrition, *Science,* 7 May 1943

As time goes on we are going to see an ever increasing use of processed foods. It is the responsibility of the food scientist and the nutritionist working as a team to insure that these foods are not only appealing to the eye and to the palate, but that they are so formulated as to properly nourish even the least endowed of our people.

—Drs. J. C. Alexander and S. J. Slinger, professors of nutrition, *Special Report No.5* (Geneva, New York: Agricultural Experiment Station, July 1971)

187

When man starts competing with nature in the blending of food elements he should be sure that his formula does not bear the skull and crossbones.
　　　　　　　—Dr. Paul B. Dunbar, Commissioner of FDA, August 1949

Food chemistry will expand and nutritional science will evolve. There is always the hope that the latter science will keep an eye on the former, eradicate the disadvantages that follow from the former, and in its turn lend advice that will confer nutritional advantage upon future reforms.
　　—James Lambert Mount, M.D., *The Food and Health of Western Man* (New York: John Wiley and Sons, 1975)

>>>>>>>>>

In 1966 J. G. Davis, former chairman of the Society of Chemical Industry Food Group in England, predicted that by A.D. 2000 it will be economically feasible to synthesize amino acids, proteins, sugars, fats, carbohydrates, and all known vitamins and flavors. Davis forecast that efficient extraction of protein from leaves and waste vegetable matter, and the making of protein by microorganisms, will be possible. By the twenty-first century, all food will be consumed from plastic and other packaging that allows for long storage. Prepared meals will be provided by vending machines. The diet will be a blend of "natural" derivatives such as leaf protein and artificial ones such as synthesized fat.

Many of Davis's predictions have already been realized. Some amino acids have been synthesized and are being used to fortify foods. Others, with adequate demand and economic reward, could be manufactured.

The nutritional value of individual protein foods of vegetable origin, especially corn, can be improved by the addition of lysine. Other foods can benefit from the addition of other amino acids, notably methionine and threonine. But food fortification with amino acids has limitations and pitfalls. If the protein content of the diet is low, the addition of a single amino acid—especially one which is added in excess—can cause an imbalance in the total amino acid

188

content of the whole diet. For example, too much lysine added to rice, when eaten by rats, will depress rather than accelerate their growth.

Researchers speculated about how amino acids linked together to form protein, the basic component of living organisms. They discovered that under appropriate conditions a mixture of fourteen amino acids could be linked together in the same manner as they are linked together in natural protein. The substance synthesized was called *protenoid* and the way in which it was produced was termed *pansynthesis*. Protenoid has many properties similar to those of natural protein. It can be split by enzymes that degrade protein. In rat studies the component amino acids in protenoid demonstrated nutritional values.

Davis's predictions about synthesizing sugar are now being fulfilled. Experimentally, cellulosic wastes are being converted by enzymatic processes into glucose and then into fructose, sorbitol, glycerol, and mannitol, as well as many other varied products. His predictions about synthesizing fats, carbohydrates, vitamins, and the development of proteins from nontraditional sources have all been discussed elsewhere in this book. To some extent, all have become realities.

Davis's comments about food vending and food automation deserve special attention. The trend is an inevitable outgrowth of a mechanical approach to food production, which values technological efficiency above all other qualities. Many agricultural processes are now automated: we use combines and mechanical harvesters, confine poultry in cages, milk dairy cows with machines, prepare feed rations by computer, inseminate cows artificially, and so on. Automation has extended through each phase of food producing, processing, storing, and distributing. Now it is apparent in the last phase as food reaches consumers.

Automation and Consumers

In the 1940s a large company in northern England attempted by means of mechanization to distribute food as efficiently as possible to employees who ate in a canteen kitchen. Soup was delivered through a valve. Even the serving of boiled cabbage was mechanized. An appropriate lever was depressed, a worm gear put into motion, and a uniform portion of the cabbage was extruded from a broad-gauged nozzle set above the level of the patron's tray. This device, though effective, hygienic, and rapid, was not esthetically pleasing and had to be abandoned at the request of the canteen's patrons. Finally, the automatic operation failed because some patrons insisted on being allowed some margin of choice in their meals.

Similar equipment is used today in restaurants where controlled portions of reconstituted mashed potatoes, flavored with a buttery taste, gush from a dispenser. On contact with air, the mass congeals on the patron's plate. The dispenser, kept in the kitchen, is not seen by the patron. Other automation in restaurants includes computerized cookstoves, computerized dispensers to measure portions of alcoholic beverages, and computerized checks.

Computerization is used extensively in institutional feeding. A trade journal described an automated cafeteria arrangement, complete with computers and conveyor belts, that eliminated interpersonal contact in food service. The article provoked someone responsible for university food services to comment on the dehumanizing aspects of food automation: "The total lack of concern for those operating the equipment and . . . the impact of automation on the customers left me feeling very sad about the future of food service. I, for one, do not plan to stumble blindly down the stainless-steel transistorized road to impersonal anonymity for the sake of alleged progress and economy. My decisions about menus and equipment changes will be strongly influenced by the effect of those changes on people."

Computerized checkouts at supermarkets are a reality in many areas, made possible by the Universal Product Code adopted by the food industry for price marking. Some supermarkets have electronic systems hooked up to banks so that patrons' checks given at the supermarket are immediately registered at the banks. Both systems have come under criticism by consumer groups.

Unmanned supermarkets already exist. In Japan one pilot unmanned supermarket contains sixty-seven vending machines, some refrigerated, displaying some three thousand food items from soups to edible seaweeds. All machines are linked to a computer. Customers insert magnetic cards into machines and punch selector switches. The computer records the sale and releases a door lock. At the checkout counter, a single cashier (as yet not replaced by automation) inserts the magnetic card into a special cash register and the computer provides an instantly itemized sales slip.

Biomedical engineers at the National Aeronautic and Space Administration (NASA) developed meals consisting of nonperishable meal packages for moon-bound astronauts. The basic meal, in a plastic bag, consists of an entrée, two side dishes, dessert, and a beverage, with a twenty-one-day menu cycle to provide variety. The astronaut kneaded the food in his hands for three minutes, then cut off the neck of the bag, put it in his mouth, and squeezed out the contents.

At present NASA is studying the use of such foods as a "meal system for the elderly" by mailing such food packages at regular intervals to shut-ins. John Keats, a social critic, found this project repellent and inhuman and "just another example of scientific bureaucracy's frequently misplaced enthusiasm." He noted the special problems of the elderly in coping with irregular deliveries, opening the plastic bags, and attempting to direct them toward the mouth. Keats added, "I believe that old people need company more than they need systems. . . . There is much more to food than nourishment and much more to a meal than food. . . . Old people, like all other people, need to eat food prepared with loving hands . . . in communion with friends. . . ."

Although some automation in daily life is agreeable and welcome, extreme forms of food automation raise many questions. They are germane, though they are beyond the scope of this book. By replacing human labor automation reduces employment—in times when we are beset by mass unemployment. Mechanization involves heavy demands on fuel and energy resources in all food systems, in agricultural practices, methods of packaging, distributing, storing and preparing—in times when we are beset by energy shortages. Do we need, and want, more and more automation, simply because it is technically feasible, at the expense of dehumanization and deculturation? Indeed, the price is high.

Will Engineered Food Solve World Hunger?

The greater part of all our food—estimated by one food scientist to be as high as 98 percent of the total caloric intake—passes through the hands of food processors. Of itself, the food industry creates practically nothing. Its function is to transform usually short-lived basic foods into palatable and attractive forms and, most important, into forms that keep for a long time. Hence, the greatest responsibility for the health of consumers rests with food processors. Yet a rough calculation in England suggested that there was only one acknowledged nutritionist to every seventeen food firms. There is no reason to believe that the situation in America is different. It becomes apparent that much food processing is being done by food chemists, food scientists, and food technologists, none of whom are primarily nutritionists. These experts, with great technological skill, have been responsible for the increasing emphasis on developing an all-synthetic food supply.

Frequently the rationale is global concern for hunger and starvation. The view expressed in a letter to the editor is typical: "The developing nations . . . are passing through very difficult times. Hunger and poverty are widespread. . . . Chemists can certainly help solve the problem by producing better fertilizers, better variety of seeds which can withstand the rigors of nature, tapping new sources of foods, and most important of all, making synthetic foods and fuels. . . ."

Many responsible scientists reject this view. Synthetic foods will not solve global hunger and starvation. Such products have to be bought before they can be eaten. The hungry have no money, and the manufacture of novel foods does not provide any increase in income for the poor. The only viable solution to the world food problem is for poor countries to increase their production of crops and

192

animals—and incomes—on millions of small farms, thus stimulating their own economies.

Food technologists view Third World countries as potential outlets for food sales, as was clearly demonstrated with infant-feeding formulas. "The profit-study picture makes the case for more foreign ventures," reported one food engineer, who added, "and one of the benefits we get from big outfits like General Foods is that the big multinationals are more able than anyone else to solve world food problems. They'll tell you this at the Harvard Business School: engineered foods are the hottest invention since fire, and if you want to make money, you better get into the engineered foods right away."

Jerry Hess, editor of *Snack Food*, while admitting that snack foods are "certainly one of the more frivolous of human foods," suggested that such items could be used as a nutritional vehicle for the world's hungry. "In rural areas where the niceties of electricity, refrigeration, etc., don't exist, many snack foods have relatively good keeping qualities. . . . They are a form of status symbol. A good example is the success Coca Cola and PepsiCo have had overseas. Being able to afford a soft drink is one of the simple pleasures in life for many. Even a soft drink is beyond the means of large numbers, however. A bag of chips or a cookie provide similar ego boosts. . . ."

Nutritional Science versus Food Technology

Studies suggest that the public makes a distinction in evaluating the outcomes of scientific work and technological work. The public views the impact of technology upon society with wariness and some skepticism.

A committee of the American Academy for the Advancement of Science, in studying the issues of scientific freedom and responsibility, found the problem "complex and formidable" in applied science and technology. Often, the committee reported, the development of innovative technology has given rise to undesirable effects. Hence, the committee rejected the notion that if something is technically possible, it should be pursued. The committee recommended that the possible repercussions of new technology should always be critically evaluated before they are introduced, and after introduction should be constantly monitored.

Some observers have argued that technology has become the source of disquieting changes in the human condition and that it is running rampant, out of control. The argument is perhaps best expressed by Jacques Ellul in his description of the "technological phenomenon" as "a pervasive situation where decision-making processes are so structured as to admit of only one outcome—the rather blind, never-ending implementation of new techniques."

In order to reverse the trend toward an all-synthetic diet, it is necessary to distinguish clearly between nutritional science, which can provide human nourishment, and food technology, which presently treats food products like any other consumer commodity.

The policies and management of food, "from garden to gullet," need to be controlled by responsible persons who understand human nourishment. That entails an entirely different concept of nourishment than is presently being applied. Soil quality is basic to crop, animal, and human nourishment and health. The nineteenth-century notion of soil as merely a mechanical support for roots must give way to twentieth-century appreciations of ecological interrelationships and the importance of biological vitality and balance in the soil. Plant and animal breeding efforts must be shifted from biological exploitation to total nourishment. Emphasis must be given to quality crops and livestock, not merely quantity yield.

Food processors need to modify their views of the food supply. The present view is that profit can be realized only by producing and promoting fabricated foods. Doubtless, profits can also be made by providing whole nourishment.

Federal agencies such as the FDA and USDA have failed in their responsibilities. They have not attempted to keep what has been dubbed "unfood" out of the marketplace, nor have they warned consumers about the dire long-range health problems that result when food choices are made primarily from such items.

The consuming public needs enlightenment. People need to understand the value of wise food selection, of choosing the traditional

194

foods that have nourished human beings through the centuries. People also need to be able to distinguish between real food and the great pretenders. People need to appreciate what Ivan Illich termed "the perils inherent in the industrialized production of food."

Appendices

Index

PRINCIPAL BOOK SOURCES

Accum, Frederick. *Death in the Pot*. London: circa 1830.

Balfour, Lady Eve. *The Living Soil and the Haughley Experiment*. New York: Universe Books, 1975.

Bicknell, Franklin, M.D. *Chemicals in Food and in Farm Produce: Their Harmful Effects*. London: Faber & Faber, 1960.

Food Colors. Washington, D.C.: National Academy of Sciences, 1971.

Hall, Ross Hume, Ph.D. *Food for Nought: The Decline in Nutrition*. New York: Vintage, 1976.

Inglett, George E., Ph.D., ed. *Fabricated Foods*. Westport, Connecticut: Avi Publishers, 1975.

Jennings, Isobel, M.D. *The Vitamins in Endocrine Metabolism*. Springfield, Illinois: Charles C. Thomas, 1970.

Mount, James Lambert, M.D. *The Food and Health of Western Man*. New York: John Wiley & Sons, 1975.

Pfeiffer, Carl C., M.D., Ph.D. *Mental and Elemental Nutrients*. New Canaan, Connecticut: Keats Publishing, 1975.

Pyke, Magnus, Ph.D. *Automation: Its Purpose and Future*. New York: Philosophical Library, 1957.

——— *Boundaries of Science, The*. Middlesex, England: Pelican-Penguin, 1961.

——— *Food and Society*. London: John Murray, 1968.

——— *Food, Chemistry, and Nutrition*. London: The Royal Institute of Chemistry Lectures, Monographs and Reports, no. 5, 1954.

——— *Food Science and Technology*, 3rd ed. London: John Murray, 1970.

——— *Man and Food*. New York: McGraw-Hill, 1970.

——— *Nothing Like Science*. London: John Murray, 1957.

——— *Science Century, The*. London: John Murray, 1967.

——— *Synthetic Food*. London: John Murray, 1970.

——— *Technological Eating, or Where Does the Fish Finger Point?* London: John Murray, 1972.

——— *Townsman's Food*. London: Turnstile Press, 1952.

Tannahill, Reay. *Food in History*. New York: Stein and Day, 1973.

PRINCIPAL FOOD TRADE JOURNAL SOURCES

ABBREVIATIONS FOR TITLES CITED

Adv. Age	*Advertising Age*
Agric. Res.	*Agricultural Research*
Am. Agric.	*American Agriculturist*
Am. Dairy Review	*American Dairy Review*
Bakery & Prod. Management	*Bakery and Production Management*
Baking Ind.	*Baking Industry*
Baking Prod. & Mktg.	*Baking Production and Marketing*
Bus. Wk.	*Business Week*
C&EN	*Chemical and Engineering News*
Chem. Progress	*Chemical Progress*
Chem. Wk.	*Chemical Week*
Community Nut. Wkly. Rept.	*Community Nutrition Weekly Report*
Consumer Bull.	*Consumer Bulletin*
Consumer Repts.	*Consumer Reports*
Consumers' Res. Mag.	*Consumers' Research Magazine*
DRINC	*Dairy Research, Inc.*
Drive-in Fast Serv.	*Drive-in Fast Service*
Farm J.	*Farm Journal*
Fed. Reg.	*Federal Register*
Fd. & Cos. Tox.	*Food and Cosmetics Toxicology*
Fd. & Drug Pkging.	*Food and Drug Packaging*
Fd. Chem. News	*Food Chemical News*
Fd. Engineer.	*Food Engineering*
Fd. Management	*Food Management*
Fd. Proc.	*Food Processing*
Fd. Prod. Develop.	*Food Product Development*
Fd. Tech.	*Food Technology*
Hlth. Bull.	*Health Bulletin*
Hlth. Fd. Retailing	*Health Food Retailing*
Inst./Vol. Fding.	*Institutions/Volume Feeding*
J. of Sc. of Fd. & Agric.	*Journal of the Science of Food and Agriculture*
Mktg. Res. Rept.	*Marketing Research Report*
Quick Frozen Fds.	*Quick Frozen Foods*
Rest. Bus.	*Restaurant Business*
Snack Fd.	*Snack Food*

NOTES

Chapter 1. THE GREAT PRETENDERS

(Liebig). More than a century later, the shortcomings of artificial fertilizers have become better recognized. Comprised of a few major elements, artificial fertilizers may lack vital nutrients, including micronutrients and as yet unidentified but critical elements. Nor do such fertilizers possess the subtle balances of nutrients to each other that are found in natural fertilizers. See W. A. Albrecht, *The Albrecht Papers;* Lady Eve Balfour, *The Living Soil and the Haughley Experiment; Qualitas Plantarum,* vol. 23, no. 4, 1974; *Scientific Papers 1054 & 1117* (Morgantown, West Virginia: Agric. Exp. Sta., 1973); *Organic & Conventional Crop Production in the Corn Belt* (St. Louis, Missouri: Center for the Biol. of Nat'l. Systems, June 1976.

(Perkins). *Food Colors.*

(Synthetic foods). Numerous laboratory experiments have been conducted with animals on semi- or all-synthetic diets. The inadequacies have been reported. E. W. H. Cruickshank, *Fd. & Nut., the Physiological Bases of Human Nut.* (Edinburgh: E & S Livingstone, 1946); also *J. Biol. Chem.,* vol. 13, 1912; ibid., vol. 23, 1915; *J. Physiol.,* vol. 44, 1912; *Science,* 10 Jan. 1941. More recently, such purified diets have been found useful for metabolic studies. Variables in foods (that may contain unknown substances with unknown effects) can be eliminated by use of purified diets. Although animals may be sustained on such diets for short-term experiments, such studies give no indication regarding the long-term effects. *Nut. Reviews,* Aug. 1965; *Am. J. of Cl. Nut.,* Jan. 1974; *S. Afr. Med. J.,* vol. 48, 1974; *Science,* 3 Nov. 1967; *Lancet,* 23 Dec. 1967; *Cancer Res.,* vol. 29, 1969. Few experiments have been conducted with humans existing on synthetic diets. The space program stimulated interest, but to date, experiments have been for relatively short periods, such as nineteen weeks, and give no indication of the adequacy or inadequacy of such diets for the entire population over a human life span or several generations. *Nature,* vol. 205, 1965.

(Wynder, Carroll). Conference on Nutrition & Cancer, sponsored jointly by Institute of Human Nutrition, and Cancer Center, Columbia University, held in New York City, November 1976.

(1972 USDA study). *Mktg. Res. Rept.* 947, Wash., D.C.: Eco. Res. Serv. USDA, Mar. 1972.

(Vocabulary). G. E. Inglett, *Fabricated Foods.*

(Continuous bread). P. J. Booras; Brit. pat. 1968.

(Veal-ham pie). Dell Fds. Eng.

(Inconvenience fds.). M. Pyke, *Technological Eating.*

CHAPTER 2. RESTRUCTURED ANIMAL PROTEIN

(Formed fish). *Quick Frozen Fds.*, May 1976.

(Fish sticks). Ibid., Oct. 1975.

(Fish fillets). Ibid., Nov. 1975; Ibid., Feb. 1976; *Fd. Management*, Dec. 1975; *Rest. Bus.*, Nov. 1975.

(Special cutter). *Canner/Packer*, Jan. 1976.

(Extruded shrimp). *Canner/Packer*, Jan. 1976.

(Crescent shrimp). Ibid., Jan. 1976.

(Shrimp cutlets). *Inst./Vol. Fding.*, 15 July 1975.

(FDA regulations). *Quick Frozen Fds.*, Mar. 1976.

(Eggs). *Dupont Mag.*, July/Aug. 1976. Machinery to reconstitute eggs consists of thirty-two rotating jacketed blocks, each containing six molds into which steel cores are inserted. Liquid eggwhite, injected around the cores, hardens when the blocks are rotated in hot water. The cores are then withdrawn and the cavities filled with egg yolk. The product is boiled, cooled, packaged, frozen, and shipped.

(Beef rolls, phosphate, salt). *Ill. Res.*, Winter 1975. Beef rolls are made by mixing chunks of beef, ground beef, salt, phosphate, water, and seasonings. The high levels of salt and phosphate may be health hazards, especially when beef rolls are served to hospital patients.

(Roast beef). *L.A. Times*, 7 Dec. 1975.

(Steak). *Fd. Proc.*, May 1975.

(Sheraton). *New York Times*, 22 Jan. 1976.

(Prices). San Francisco *Chronicle*, 23 May 1976.

(Consumer reaction). *J. of Fd. Sc.*, vol. 41, 1976.

(Grill steak). *Rest. Bus.*, Apr. 1976. Reconstituted steak is made from meat that has been flaked. Then spices, additives, or flavor enhancers are added, and the blend is held together by hydraulic pressure. The product can be produced wafer thin or up to two and one fourth inches thick. Such "steaks" are frozen, packaged, vacuum sealed, boxed for storage, and shipped to feeding institutions.

(Veal grill). *Quick Frozen Fds.*, Jan. 1976; also ibid., Feb. 1976.

(Lamb chops). Ibid., Mar. 1976.

(Pork). *Fd. Proc.*, July 1974; also *Inst./Vol. Fding.*, 1 Feb. 1976.

(Extended meat). *Fd. Proc.*, May 1976.

(MDM-export & pet fd.). Ibid., Feb. 1975.

(Guidelines). *MPI Bull.* 865, Washington, D.C.: USDA, 6 Nov. 1974.

(Rules relaxed). Newark [N.J.] *Star Ledger*, 15 Aug. 1976.

(Heat pasteurization). *Sc. in Agric.*, Univ. Pk., Pa.: Agric. Exp. Sta., Spring 1976. Heat treatment of MDM is tricky. If the heat is too high or too prolonged, MDM's capacities are impaired to emulsify, disperse fat, and hold water. Spoilage caused by microbial growth is a major problem in MDM that is not frozen, whereas oxidative rancidity of the fat portion is a major cause of deterioration in frozen MDM.

(Fat in MDM). Wash. *Post*, 27 May 1976.

(Garbage). Boston *Globe*, 1 Sept. 1976.

Notes

(Mayer). *Fd. Proc.*, Oct. 1976.

(SPF). Ibid., July 1975.

(TFGB). USDA release, 5 Oct. 1977.

(Chicken legs & fish fillets). *Quick Frozen Fds.*, May 1976. SPF, combined with MDM, is stuffed into casings and heated to form firm meat products that are sliced or diced. Or the resulting products are used as bases for formulating other products.

(CBTS). *Fd. Proc.*, Apr. 1975. To make CBTS, raw beef fat is ground, heated, disintegrated, and then centrifuged to separate the solids from the liquid (tallow). The solids, which contain up to 25 percent protein and up to 6 percent fat, are sold as CBTS, either frozen or in a shelf-stable dried form.

(Mich. court). Ibid., Oct. 1974.

CHAPTER 3. ANIMAL PROTEINS VERSUS TEXTURED VEGETABLE PROTEINS

(Kellogg). *Hlth. Fd. Retail.*, Mar. 1972.

(Unpopularity). *Cornell Hotel & Rest. Quart.*, Aug. 1966. In a report about food in Japanese prison camps in World War II: "Many attempts were made to make soybeans palatable and digestible. The only satisfactory method [proved] to be one in Indonesia, involving inoculation with a fungus. Otherwise, these beans were liable to give rise to much digestive disturbance when used in any quantity, even if first reduced to a fine meal." *Nature*, 19 Jan. 1952.

(Epithet). *Nut. Today*, Mar.–Apr. 1973.

(Spun fibers). Two basic methods, extruding and spinning, produce Engineered Protein Products. By extruding, a mixture of defatted soy flour or flakes and water is forced through a cylinder under increasing pressure. The combined heat and pressure breaks down the protein's cellular structure. It produces a porous, fibrous mass with the consistency of cooked meat. In spinning, defatted soybean flour is dissolved in an alkaline liquid and forced through a spinning die pierced with thousands of tiny holes. The emerging liquid jets congeal in an acid bath into separate threads of protein that can be textured to resemble just about any meat product: slices, flakes, cubes, or granules. The process is used mainly to produce meat analogs. Such processings, with high heat and pressure, denature the protein in the original food. Hexane, a toxic solvent, may be used in processing to remove the lipids (fatlike substances) from the soy flakes, and it remains as a residue. Such processings also require high energy use, a factor that should be considered with all highly processed foods.

(Annual production). In 1973, annual sales for vegetable proteins were $83 million. By the end of the decade, sales are expected to be as high as $1.5 billion. *Fd. Engineer.*, Nov..1973. By 1980 half of the chopped beef patties, hot dogs, sausages, and other processed meats consumed will be supplied by vegetable protein. *Fd. Proc.*, July 1974. The frozen-food market predicts $100 million sales by 1980 for frozen soy extenders and analogs. *Quick Frozen Fds.*, Oct 1975. Spun vegetable proteins, which are noncellular, can be frozen and thawed many times, with little change in texture, flavor, or appearance. *The Many Faces of Engineered Protein Products* (Chicago, Illinois: Nat'l. Live Stock & Meat Brd., Aug. 1974).

(Alternate protein fds.). Protein foods from cottonseed, peanuts, rice, rapeseed, and field beans are being investigated for human or livestock consumption. Attempts are being made to extract protein from sesame and sunflower seeds, alfalfa, grass, tree leaves, and other land and water plants. Proteins and protein concentrate from animal and dairy by-products and from fish, including products from skim milk, whey, brewer's and baker's yeasts, animal blood, and tannery wastes are being utilized or explored. Less traditional sources include Single-Cell Protein (SCP), a generic term for crude or refined sources of protein from unicellular or simple multicellular organisms such as bacteria, yeast, fungi, algae, and protozoa. SCP has been obtained by using as a growing medium sewage, garbage, wood pulp, or petroleum. Yeast, produced from hydrocarbons, which are by-products of petroleum refining, is being studied for animal feed. None of these substances are of high biological quality equal to animal protein sources. Some of them contain antinutritional factors that require deactivation; some contain poisons that require special processing; some may contain toxic contaminants. SCP is high in nucleic acids, which can be hazardous.

(Isolates). The use and promotion of expensive isolated protein concentrates by food processors as a nutritional savings was criticized by Dr. George M. Briggs, who pointed out that isolated textured protein concentrates are more expensive than the original crude proteins existing in the natural food source of the protein. Briggs, "Nutritional Aspects of Fabricated Foods," in *Fabricated Foods.*

(Dietician's observation). *FDA Consumer*, Apr. 1975.

(Simulated meats). *Nat'l. Observer*, 14 Apr. 1973. Increasingly, soy products are used in restaurants and other institutions and are difficult to avoid. The soybean and soy products may be allergens for individuals.

(Food distributor not selling blend). Easton [Pa.] *Express*, 12 June 1973.

(Extender limited). *Consumers' Res. Mag.*, Feb. 1974.

(Fat limited). *Consumer Repts.*, Feb. 1975. The use of soy extenders has been a boon to processed meat from cheaper cuts. The result, described by a soy product supplier, is a product that is "firm, juicy and full of bounce." *Lancet*, 11 Nov. 1972.

(Costly). *Progressive*, Sept. 1975.

(Diabetics). *FDA Consumer*, Apr. 1975.

(Terms). *Consumer Protection*, State Dept. of Agric., Oregon, May 1973; also *Media & Consumer*, Dec. 1973; *Adv. Age*, 9 Apr. 1973.

(Truth in menu). *Rest. Bus.*, May 1976.

(Fast-food illegalities). *Media & Consumer*, Jan. 1974.

(Institutional usage). *Nat'l. Observer*, 14 Apr. 1973.

(Recipes). *Fd. Proc.*, Nov. 1974.

(Total replacer). Ibid., Oct. 1976.

(Analogs costly). *Good Fd.*, Nov. 1973; also Phila. *Eve. Bull.*, 24 Nov. 1972.

(Eng. chicken, trout). *Lancet*, op. cit.; also *County Agent Notes*, N.H. Coop. Ext. Serv., 31 Jan. 1973.

Notes

(Untrue claims). *Hlth. Fd. Retailing*, Mar. 1972; also press release, Arco Pub., 29 Nov. 1972.

(Nut'al. evaluation). *Professional Nutritionist*, Summer 1974; also *J. of Fd. Sc.*, vol. 38, 1973; *J. of Amer. Diet. Assoc.*, vol. 63, Nov. 1973.

(Alkali treated). *Nut. Reviews*, Aug. 1973.

(Amino acids). *Mktg. Res. Rept.* 947, USDA, Mar. 1972. Soy protein isolate, as contrasted to soy flour and soy protein concentrate, has a lower content of the sulfur-containing amino acids methionine and cystine. In soy protein, methionine and cystine are somewhat lower than in whole egg protein. The biological value for full-fat soy flour is lower than for defatted flour, possibly due to the heat processing. There is a wide variation of protein efficiency ratio (PER) in soy protein isolates. The amino acid composition of a protein does not always reveal the extent to which a particular amino acid may be available or utilized by the body. L. D. Williams et al., *Nutritional Value of Soy Protein Products in Fds.* (Wash., D.C.: League for Internat'l. Fd. Ed., undated).

(Fat & non-nutrients). *Lancet*, op. cit.

(Rehydrated). *The Many Faces of Engineered Protein Products*, op. cit.

(Sodium). Washington *Post*, 30 Jan. 1975.

(Oxalic & phytic acids). *J. Nut.*, May 1972. Folic acid, zinc, potassium, and other nutrients should be added to textured vegetable protein products used in the USDA's child-feeding programs, suggested Mead Johnson & Company, noting that the three first-named nutrients were among those usually provided by meat, poultry, or fish. The company noted that the USDA's limit of textured vegetable protein, combined with animal proteins, "might not be a serious problem." But these same nutritional standards may well be applied to textured vegetable protein products that are used in greater amounts in the diet with little or no animal proteins. In such cases, serious nutritional inadequacies could result if these nutrients were omitted. *Fd. Chem. News*, 28 May 1973. A higher percentage of textured vegetable protein may be used by commercial interests than the maximum 30 percent the USDA permits in child feeding programs. Soy protein isolates can be used to replace up to 60 percent meat protein in emulsified meat systems to form nonstandardized products. *Fd. Proc.*, Apr. 1975.

(Selenium). *Nut. Notes*, July 1975.

(Products precede regulations). *The Many Faces of Engineered Protein Products*, op. cit.

(Appraisal). *Lancet*, op. cit.

(Antinutritional factors). *Toxicants Occurring Naturally in Fds.*, Pub. 1354, Washington, D.C.: NAS/NRC, 1966.

(Vaupel). *New York Times*, 15 May 1973. The USDA stated that soy products such as soy milk are lower in food value than cows' milk. Although the protein content is somewhat similar, the soy protein is not utilized in the human as well. *Home & Garden Bull.* 208, USDA, June 1974. In 1976, the FTC seized a soy protein mix intended as an infant formula, claiming that the advertisement falsely represented soy milk as nutritionally adequate in normal growth and development in infants. It required that the company place a warning on the product's label: "not for use in diets of infants under one year of age unless recommended by a physician."

(Imbalance, deficiency). *The Many Faces of Engineered Protein Products*, op. cit. (Fd. Res. & Action Center). Statement Submitted to USDA Protesting Use of Alternate Fds. in Sch. Fd. Serv., May 1973. The statement also condemned the USDA's approval of a fortified breakfast cake, which was described by the center as a "substandard piece of confection propagandized as an all-purpose nutritional diet by a multimillion-dollar conglomerate" and enriched macaroni "disguised as a starch."

CHAPTER 4. BIDDY'S SHELL EGGS VERSUS EGG SUBSTITUTES

(Early fabricated eggs). M. Pyke, *Townsman's Food.*
(1972 simulated egg). *New York Times*, 12 Feb. 1972.
(Bakers). *Fd. Proc.*, Nov. 1974; also ibid., Apr. 1976; *Baking Ind.*, Apr. 1976.
(Cholesterol). *Fd. Proc.*, July 1976.
(Advertisement). *New York Times Mag.*, 17 June 1973. "What does Standard Brands do with all those yolks removed from Egg Beaters? 'Various things,' says a company spokesman. 'Some go into bakery products, some into cosmetics, some into pet food.' 'Toward better-fed dogs,' snorts Texas' Dr. [Roger J.] Williams." *Med. Wrld. News*, 15 Mar. 1974. "Natural egg-type flavor" is offered to food processors for use in products. The advertisement announces, "It's the right breakthrough at the right time. Using all natural ingredients, we have created a completely convincing egg flavor. With physicians warning of the high cholesterol content of real eggs, substitutes are in great demand." The product is intended for use in new food concepts, no-egg foods such as scrambled eggs, mayonnaise, French toast, eggnog, pancake batter, cookies, batter for deep-fat frying, cake, and salad dressings. *Fd. Proc.*, June 1975; also ibid., Aug. 1975.

(Taste). *Moneysworth*, 6 Jan. 1975; also *Consumer Repts.*, Mar. 1974.
(Inaccurate claims). Ibid., op. cit. Egg substitutes are not replacers for "farm fresh eggs." A cholesterol-free substitute for an animal protein should possess the same nutritional value as the cholesterol-containing natural product. A researcher recognized this in his study of egg yolk. He used casein as a protein source and added cystine at various levels in order to balance out the amino acid content of milk versus egg white protein. However, the amino acid levels in egg white and casein are not identical, and therefore the use of egg instead of milk protein may have made a difference in serum cholesterol response. F. A. Kummerow, *The Role of Eggs in the Amer. Diet*, Urbana, Illinois: Burnsides Res. Lab., Univ. of Ill., undated.

(Trace mineral deficiency). C. C. Pfeiffer, *Mental & Elemental Nutrients.*
(B vitamins). *Consumer Repts.*, op. cit. The problem of nutritional equivalency between fabricated and traditional foods is one of growing concern to the Canadian Food Directorate, Health Protection Branch, Canada, the counterpart of our federal FDA. D. G. Chapman, assistant director-general, announced in 1975 a proposed new regu-

Notes

lation to deal with the nutritional requirements for products simulating whole eggs. Such products must contain a specific quantity and quality of protein, as well as specified vitamins and minerals, and not more than 3 milligrams of cholesterol. The regulations controlling the nutritional value of substitute eggs, as well as meats, are just the beginning, with an anticipated series of regulations of other substitute food products that have appeared or are likely to be appearing in food stores. D. C. Chapman, *Canadian Legislation & Regulations Affecting Fd. Quality & Safety*, paper, Am. Chem. Soc., 26 Aug. 1975.

    (Carbohydrate). *Consumer Repts.*, op. cit.

    (Sodium). Ibid., op. cit.

    (Navidi & Kummerow). *Peds.*, Apr. 1974; also *J. of Fd. Sc.*, Jan.–Feb. 1975. Additional rat studies with egg substitute products showed that animal growth was better with whole egg than egg substitute either with or without vitamin and mineral supplementation. The researchers cautioned: "Care in the formulation of the egg substitute product by the food industry to include essential nutrients normally obtained in whole eggs should be considered, or adequate information regarding the deficiency of these nutrients in relation to whole egg should be made available." *J. of Amer. Diet. Assoc.*, Mar. 1976. Another animal-feeding experiment reinforced the findings of Navidi and Kummerow. Using white leghorn chicks, university students tested several commonly used food products, including egg substitute products, breakfast cereals, and doughnuts, compared with hard-cooked eggs and starter mash. Each food was fed for a period of three weeks, then all chicks were shifted to a starter mash. Chicks receiving the Egg Beaters® lost weight, and all of the animals in this group were dead within twelve days after the trial began. The only groups without mortality were those on hard-cooked eggs and the starter mash diets. Experiments are criticized in which animals are fed single food items. But the students commented, "It is obvious that one does not consume a single food. However, single food feeding dramatically demonstrates biological differences better than comparing them in food analysis tables." *Poultry Times*, 26 Mar. 1975.

CHAPTER 5. OLD BOSSIE'S CREAM VERSUS IMITATION CREAM PRODUCTS

    (Anderson). *Farm J.*, Dec. 1974.

    (Coffee creamers—history). *L.A. Times*, 28 Aug. 1972.

    (Wisconsin decision). *Quick Frozen Fds.*, Jan. 1976; also *Drive-in Fast Serv.*, Jan. 1976.

    (Annual volume—nondairy creamers). *Am. Dairy Review*, May 1974.

    (Advertising claims). *Rest. Bus.*, Aug. 1974; also ibid., Sept. 1975; *Fd. Management*, Mar. 1976.

    (Use with cereal). Washington *Post*, 22 May 1975.

    (Cooking ingred.). Miami *Herald*, 15 Nov. 1973.

    (40 percent market loss). *Farm J.*, op. cit.

(Components). *Inst./Vol. Fding.*, 15 Nov. 1975; also *Fd. Management*, Nov. 1975. New frozen nondairy creamers have been formulated especially for the hospital and nursing home feeding market, catering to those who must have milk-free diets. Such products are labeled "no milk," "milk fat," "milk proteins," or "milk sugar," and are suggested for use with coffee, cereal, and fruit. Soy isolate may replace sodium casein-ate in such products. Only about half as much soy isolate is needed, and it is also slightly cheaper than sodium caseinate. Hence, it reduces manufacturing cost. *Fd. Proc.*, June 1976.

(FDA—sodium caseinate). *Consumer Repts.*, Mar. 1975.

(Mayer). Washington *Post*, op. cit.

(Imitation milk). At present, imitation milk is being marketed by two companies. By 1980 it is projected to account for up to 10 percent of the fluid milk market. *Progressive*, Sept. 1975; also *New Republic*, 10 Aug. 1968.

(Nutritional equivalency). *Am. J. of Cl. Nut.*, Apr. 1969.

(Nutrition-filled & imitation milk). *Dairy Council Digest*, Mar.–Apr. 1968; also *Am. J. of Cl. Nut.*, Feb. 1969.

(Topping). *Quick Frozen Fds.*, Dec. 1975; also M. Pyke, *Synthetic Fd.*; *Quick Frozen Fds.*, Nov. 1975; *Farm J.*, op. cit.

(Advertisements). *Rest. Bus.*, June 1974; also ibid., Apr. 1976; Ibid., Apr. 1974; *Fd. Proc.*, Nov. 1975; *Rest. Bus.*, May 1976; *Drive-in Fast Serv.*, Feb. 1976; *Fd. Proc.*, Nov. 1975.

(Formulation). M. Glicksman, "Carbohydrates for Fabricated Fds., Dairy Analogs," *Fabricated Fds.*

(Imitation sour cream). Ibid.

(Mellorine). S. H. Marple, "Milk & Its Substitutes," N. H. Consumer Competency Leaflet 6 (Durham, New Hampshire: Univ. of N.H., Nov. 1970); also *Am. Dairy Review*, Oct. 1975; "Ice Cream & Similarly Frozen Foods," Information sheet, Chicago, Illinois: National Dairy Council, 1971.

CHAPTER 6. IMITATION CHEESES REPLACE NATURAL CHEESES

(Italian cheese scandal). R. Tannahill, *Food in History.*

(Cost of cheddar). *Snack Fd.*, Jan. 1976.

(Cheese scarcities). *Fd. Proc.*, Feb. 1976.

(Composition of imitation). *Fd. Prod. Devel.*, vol. 4, 1970.

(Three groups). *Fd. Proc.*, Feb. 1976.

(Total replacer). Ibid., July 1976.

(Advertisements). *Snack Fd.*, Aug. 1974; also *Fd. Proc.*, June 1974; ibid., Oct. 1975; ibid., June 1976.

(Guidelines). *Snack Fd.*, Mar. 1976. Labeling of artificial cheese flavors differs from country to country. In Germany, though such flavors are regarded as artificial, in accordance with essence regulations, such flavorings do not have to be declared on the label. In England and Switzerland, the term *Nature-Identified* may be used on the

Notes

label if the artificial cheese flavors contain substances that are identical in chemical structure to natural cheese products. This term is unacceptable in the United States. In France, *artificial* must appear on the label. *Fd. Proc.*, May 1976.

(Golana). *Am. Dairy Review*, Sept. 1975.

(Nutritive value). *Dairy Council Digest*, May–June 1975. The nutritional comparisons of natural cheese replacers are available for a few of those products. Based on 100 gram portions, low-moisture part-skim natural mozzarella has 48 percent water, 290 calories, 27 grams protein, 18 grams fat, 3.6 grams carbohydrate, 710 milligrams calcium, 360 I.U. Vitamin A, .3 milligrams riboflavin. Mozzarella-style replacer has 47 percent water, 310 calories, 23 grams protein, 23 grams fat, 1.3 grams carbohydrate, 650 milligrams calcium, 1,200 I.U. Vitamin A, .7 milligrams riboflavin. Thus, the natural cheese contains more protein and calcium, and less fat, than the replacer. *Fd. Proc.*, June 1976.

(WARF study). *Am. Dairy Review*, June 1976.

(Artificial cheese flavors). Ibid., Jan 1976; also ibid., May 1975; *Snack Fd.*, June 1974; *Fd. Proc.*, Dec. 1975; ibid., May 1974; ibid., Aug. 1975; *Snack Fd.*, Feb. 1976; ibid., Apr. 1976; *Am. Dairy Review*, Mar. 1976; *Fd. Proc.*, Apr. 1976; Ibid., Mar. 1976; *Am. Dairy Review*, Dec. 1974; *Canner/Packer*, Jan. 1976; *Rest. Bus.*, Nov. 1975; *Fd. Proc.*, Nov. 1975; *Snack Fd.*, May 1975; *Fd. Proc.*, May 1975.

CHAPTER 7. BUTTER VERSUS MARGARINE

(Questionnaire). *Am. Dairy Review*, Dec. 1975.

(Price increase). Ibid., Nov. 1974.

(More costly). *Farm J.*, Dec. 1974.

(Taste). *Consumer Repts.*, May 1963.

(Middle-class alternative). R. Tannahill, *Food in History*.

(Butter in 1800s). F. Accum, *Death in the Pot*. This rare book, lacking publication information, was probably printed in London about 1830. Accum's earlier classic, *A Treatise on Adulteration of Food & Culinary Poisons*, had been published in London in 1820.

(Chevreul). *Encyclopaedia Britannica* (Chicago, Illinois: Encyclopaedia Britannica, 1948).

(Mège-Mouriés). B. A. Fox & A. G. Cameron, *A Chemical Approach to Fd. & Nut.* (London: Univ. of London, 1973); also J. C. Drummond & A. Wilbraham, *The Englishman's Fd.* London: Jonathan Cape, 1964; R. Tannahill, op. cit.

(Butterine). Ibid.

(Hydrogenation). Technological advances made it possible to consider a wider variety of fats for margarine manufacture. The hydrogenation process made it possible to convert fats or oils from almost any source into margarine production, with any desired consistency. M. Pyke, *Technological Eating*. Coconut oil, frequently used as a constituent in margarine, is more saturated than butterfat in butter. On a weight percentage, coconut oil contains 91.2 total saturated fat, contrasted to 70.6 in butterfat from cows' milk; coconut oil contains only 7.9 total unsaturated fat, compared to 29.4 in butterfat. D. Swern, ed., *Bailey's Industrial Oil & Fat Products* (New York: Interscience, 1964). Moreover, some commercial margarines, advertised as being high in polyunsaturates from corn oil, did not upon analysis live up to their claims. Several ordinary margarines, with no special advertising claims regarding fat constituents, contained more polyunsaturates than products with special labels, and at higher prices. *Canadian Med. Assoc. J.*, vol. 93, 1965.

(Early unfavorable associations). In order to overcome consumer prejudice against margarine, a leading home economist, Ida Bailey Allen, was employed. In each of six cities a luncheon was arranged to which twenty-five to thirty of the city's key women were invited. All the food served was prepared with margarine, and it was also supplied instead of butter along with the bread. After the luncheon, Mrs. Allen arranged for a demonstration and explained the uses of the product before audiences ranging from five hundred to one thousand five hundred local housewives. J. F. Schlink, *Eat, Drink & Be Wary* (New York: Covici Friede, 1935). In the 1970s, similar tactics were used to promote margarine, this time for its purported heart benefits. In 1972 Fleischmann advertised that three thousand five hundred cardiologists ate a meal prepared especially with Fleischmann margarine. In actual fact, at no one time did more than 478 of these physicians and their guests ever sit down at a single meal to eat this particular product. Furthermore, a medical journal reported that many of the cardiologists who had attended the meeting in question escaped to a local coffee shop for hamburgers and milk shakes. E. R. Pinckney & C. Pinckney, *The Cholesterol Controversy* (Los Angeles: Sherbourne Press, 1973).

(Motivational researchers). E. Dichter, *Handbk. of Consumer Motivation* (New York: McGraw-Hill, 1964).

(1958 sales). *Chem. Wk.*, 19 Apr. 1958.

(Ads). *Consumer Repts.*, May 1963. Advertising claims for margarine were so blatant that Charles Edwards, then commissioner of the FDA, said that if drugs were advertised with such extravagant claims, the agency "would regard the products as misbranded" and would seize them. Although the FTC, the agency responsible for regulating advertisements, has made a feeble attempt at controlling margarine ads, more vigorous action is needed. Margarine ads continue to use such captions as "eat to your heart's content" with a picture of a heart wrapped in a label of the margarine product; "the change will do your heart good," "enjoy it to your heart's content," "you can aid the fight against heart disease by starting in your kitchen," etc. In 1973 the Radio Code Board of the National Association of Broadcasters attempted to bar specific health benefits claimed for margarines as well as for vegetable oils.

(Ads—physicians). For years, margarine manufacturers have attempted to influence physicians to recommend their products to patients, by means of advertisements in medical journals and by free literature and displays at medical meetings. *Nut. Today*, July–Aug. 1973. Professionals have begun to warn the public. Dr. Norman Lasser, addressing the New Jersey Academy of Family Physicians in March 1976, spoke critically about margarines and other nondairy substitutes, charging that many

210

Notes

companies are misleading consumers with low cholesterol and health claims while "they are thinking of their pocketbooks, not your arteries when they make these foods." Newark [N.J.] *Star Ledger,* 14 Mar. 1976. In research conducted at the USDA's Agricultural Research Center, hardening of the arteries in rabbits was not influenced by whether the fat in the diet of the animals was saturated or unsaturated. Press release, USDA, 15 Apr. 1975. Kummerow found that the serum cholesterol level in humans did not change significantly either after five or twenty-four hours, or after a time period up to fifty-four days of continued, moderate consumption of cholesterol-rich food. F. Kummerow et al., *The Influence of Cholesterol-Rich Fds. on the Serum Cholesterol Level of Human Subjects* (Urbana, Illinois: Burnsides Res. Lab., Univ. of Ill., undated). High levels of polyunsaturated fat in the diet may be risky. Some reports have linked cholesterol-lowering diets, high in polyunsaturates, to gallstone formation and various types of cancer. E. W. Speckman, "Coronary Heart Disease: Risk Factors & the Diet Debate." Senate Select Comm. on Nut. & Human Needs, 26 July 1974. Rats fed natural saturated fats, with some cholesterol, grew bigger and better looking, and lived longer, than rats eating unsaturated fats and no cholesterol. *C&EN,* 10 Sept. 1973. Human malignant melanomas were reported in five cases in which all patients had changed from butter to polyunsaturated margarine, and they had all changed to polyunsaturated oils or margarine for cooking for a period of time before their lesions were observed. They had all enthusiastically followed the "polyunsaturated routine." *Med. J. of Australia,* 18 May 1974. Increased consumption of polyunsaturates steps up the need for greater consumption of Vitamin E. For an in-depth discussion of how the public has been misled regarding margarines, polyunsaturated oils, cholesterol, and coronary diseases, see E. R. Pinckney & C. Pinckney, *The Cholesterol Controversy,* op. cit.

(*Trans* vs. *cis*: altered metabolism). *J. of Nut.,* vol. 102, 1972; also R. H. Hall, *Food for Nought.* Although it is generally regarded that *trans* isomers are unnatural, there is a minority view that these forms do appear in nature. A Professor Kaufmann of Münster, Germany, demonstrated that in summer, cows consume many plants from which *trans* acids are secreted into the milk and find their way in substantial amounts into butter. Microbial hydrogenation of unsaturated fatty acids within the cow also contributes a share of *trans* acids to butter. *Fd. & Cos. Tox.,* Dec. 1963. Also, some experiments showed that the effect of a hydrogenated fat on blood lipid levels is determined by its fatty acid composition, and this effect is not altered by the isomeric form of the unsaturated acids. *Am. J. of Cl. Nut.,* July 1975. In nature, the molecules of the various unsaturated acids such as linolenic, linoleic, and oleic acids are doubled back on themselves in a particular configuration. The same chemical combinations produced in the course of hydrogenation may possess molecules that are folded differently. This affects the hardness of the oil or fat produced and thus its nutritional value. Dr. Pyke, who generally expresses admiration for the achievements of fabricated foods, nevertheless cautions that nutritional alterations, through hydrogenation of fats, is a subject "still being investigated. Although the conclusions are not yet clear, it is a matter that must be borne in mind by food technologists." M. Pyke, *Fd.*

Sc. & Tech. Others concur and recommend that the effects of *trans* isomers on overall health need further investigation. *Am. J. of Cl. Nut.*, vol. 23, 1970. A good discussion of *trans* fatty acids is in *Dairy Council Digest*, Nov.–Dec. 1974.

(*Lancet* warning). *Lancet*, vol. 2, 1956.

(Sinclair). *Drug Trade News*, 1 July 1957.

(Bicknell). F. Bicknell, *Chem. in Fd. & in Farm Produce* (London: Faber & Faber, 1960).

(Kummerow—*trans* isomers). F. A. Kummerow et al., *Swine as an Animal Model in Studies on Atherosclerosis* (Urbana, Illinois: Burnsides Res. Lab., Univ. of Ill., 1974).

(Percent of *trans* acids in margarines). *J. of Am. Oil Chem. Soc.*, vol. 25, 1968.

(Kummerow). F. A. Kummerow, *The Role of Eggs in the Am. Diet* (Urbana, Illinois: Burnsides Res. Lab., Univ. of Ill., undated).

CHAPTER 8. RESTRUCTURED FRUITS AND VEGETABLES

(Fd. pellets). *Fd. Proc.*, Jan. 1976.

(Puff explosion). *DRINC*, June 1973.

(USDA res.). *Service*, July 1964.

(Puffed carrots). *Fd. Proc.*, May 1975.

(Granules, flakes). Ibid., Oct. 1976; also *Baking Ind.*, Sept. 1976.

(Shredded, bagged food). *Fd. Proc.*, Aug. 1976.

(Dehydrated food). Ibid., July 1976; also Oct. 1977.

(Onion). *Quick Frozen Fds.*, Nov. 1975; also ibid., Dec. 1975; *Rest. Bus.*, June 1974; ibid., Jan 1976; *Fd. Proc.*, Nov. 1975; *Rest. Bus.*, Apr. 1974.

(Label). *Canner/Packer*, Jan. 1976.

(Restructured french fries). Len Fredrick, *Fast Food Gets an "A" in School Lunch*. Boston: Cahner's Bks. Internat'l, 1977.

(Potato—Crum). *Time*, 8 Dec. 1975.

(Small companies). *New York Times*, 3 Nov. 1975.

(Increased cost). *Snack Fd.*, Dec. 1975.

(Steamroller). Ibid., Jan. 1976.

(Counteradvertisement). Ibid., Apr. 1976.

(PCII). Ibid., Sept. 1975.

(FDA definition). Ibid., Jan. 1976.

(Mislabeling). Ibid., Mar. 1976.

(Editor). Ibid., Sept. 1975.

(FDA decision). Ibid., Jan. 1976.

(3 yrs.). Ibid., Mar. 1976.

(PCII action). Ibid., Jan. 1976.

(Editor). Ibid., Sept. 1975.

(Refusal to rescind). Ibid., May 1976.

Notes

CHAPTER 9. REAL VERSUS IMITATION FRUITS
AND VEGETABLES

(Synthetic cherry). *Fd. Proc.*, vol. 22, 1961. The alginate-based process for novelty foods can be used not only for fabricated cherries and blueberries but also for fabricated caviar, potato chips, other vegetables, imitation low-calorie spaghetti and spaghetti sauce, meatballs and shrimp, meat and gravy loaves such as sloppy joe and barbecued meat, sausage casings, gelatinized beer, and soft drink gels. G. E. Inglett, *Fabricated Fds.*

(Gen'l. Fds. pat.). *Fd. Manufacture*, Oct. 1968.

(Pyke). M. Pyke, *Syn. Fd.*

(Raisin-flavored granules). *Fd. Proc.*, Mar. 1977.

(Imitation tomato solids & paste). G. E. Inglett, *Fabricated Fds.*; also *Fd. Proc.*, May 1974; ibid., Nov. 1974; ibid., May 1975; ibid., Dec. 1975; ibid., Dec. 1974.

(Fickle tomato). *Fd. Prod. Develop.*, Feb. 1974.

(Orange juice). Mount Vernon [N.Y.] *Daily Argus*, 3 Jan. 1973.

(Knauer). *Nat'l. Observer*, Dec. 11, 1971.

(Consumption data). "Synthetics & Substitutes for Agricultural Products, Projections for 1980," *Mktg. Res. Rept.* 947, USDA, Mar. 1972.

(Ads). *Wall St. J.*, 3 Sept. 1968.

(Fla. Citrus Comm.). *Hlth. Bull.*, 6 Mar. 1965. This 156-page report was written by Drs. Cortez F. Enloe, Jr., Frederick Stare, and Willard Krehl.

(Counteradvertisements). *Consumer Bull.*, Aug. 1964.

(Kirk). Press release, Fla. Citrus Comm., 25 July 1969.

(Orange Plus). *Nat'l. Observer*, op. cit.

(Standards of Iden.). *Pontiac Press*, 17 June 1971.

(Composition of real orange juice). *New York Times*, 7 Mar. 1975. This article included a composite chart comparing prices and nutritional values of real orange juices and synthetics, from sources provided by the USDA and processors. A powdered synthetic orange substitute for real orange juice was termed "the height of artificiality" by Elizabeth Chant Robertson, M.D., Ph.D. She wrote that it is true that the product "contains the 40 milligrams ascorbic acid per heaping teaspoon as advertised, and its content of vitamin A is higher than that in orange juice. However, orange juice has minor but useful amounts of many other substances and for that reason I believe it is preferable to such synthetic products." *Canadian Hospitals*, vol. 8, 1962.

(Lemonade). *Rest. Bus.*, Apr. 1976.

CHAPTER 10. THE STAFF OF LIFE VERSUS
CARBOHYDRATE SUBSTITUTES

(Synthetic carbohydrates). M. Pyke, *Syn. Fds.*

(Microcrystalline cellulose). *Life*, 2 June 1961; also *Fd. Proc.*, Nov. 1960.

(Burros). *Washington Post*, 28 Oct. 1976.

(Pollulan). *C&EN*, 24 Dec. 1973.

(Xanthan gum). *Canner/Packer*, Jan. 1976; also *C&EN*, 1 Dec. 1975; also *Fd. Proc.*, Jan. 1976; *Fd. Tech.*, June 1974; ibid., May 1974.

(Chicken feathers). *The Express*, Easton, Pa., 26 Aug. 1976.

(Dough conditioners). *Fd. Proc.*, Nov. 1975; also *Chemmunique*, vol. 24, 1975.

(Milk simulators). *Baking Ind.*, Apr. 1976; also *Fd. Proc.*, July 1975; ibid., June 1976.

Chapter 11. CREATIONS BY THE FOOD FLAVORISTS

(1500 flavoring ingreds.). *FDA Consumer*, Apr. 1975.

(20 Am. Co.). *Chr. Sc. Monitor*, Apr. 30, 1975.

(Analytical tools). *Chem. Progress*, Sept.–Oct. 1974.

(Bacon flavors). *Fd. Proc.*, Oct. 1975; also *Snack Fd.*, Aug. 1975; Ibid., Mar. 1974.

(Strawberry jam flavor). *Fd. Proc.*, July 1976.

(Tomato flavors). *Snack Fd.*, July 1976; also ibid., Dec. 1975.

(Onion flavors). Ibid., Mar. 1974; also *Rest. Bus.*, Aug. 1975.

(Beef flavors). *Fd. Proc.*, July 1975.

(Chicken flavors). *Fd. Prod. Develop.*, Mar. 1974.

(Flavors for soups, gravies, casseroles). *Fd. Proc.*, Mar. 1976; also ibid., May 1974; ibid., Aug. 1975; ibid., July 1975; ibid., Dec. 1975; *Inst./Vol. Fding.*, Dec. 1, 1975; *Fd. Proc.*, Apr. 1976.

(Seafood flavor). Ibid., Sept. 1975.

(Anchovy, tuna flavors). Ibid., June 1976.

(Frankfurter flavor). Ibid., May 1975.

(Ethnic flavors). Ibid., Sept. 1975.

(Taco flavor). Ibid., July 1976.

(Fat flavor). Ibid., Oct. 1976.

(Mayonnaise flavor). Ibid., Sept. 1975.

(Coconut flavor). Ibid., Aug. 1974.

(Peanut flavor). *Snack Fd.*, June 1975; also ibid., May 1974; *Baking Ind.*, Dec. 1975; *Fd. Proc.*, Sept. 1975.

(Custard flavor). *Snack Fd.*, May 1975.

(Tomato-bacon flavor). Ibid., Feb. 1975.

(Pasteurized milk flavor). *Chem. Progress*, May 1969.

(Injected flavors). *Here's Hlth.*, May 1973.

(Sales scent). Quoted by Sen. Norris Cotton, N.H., Newsletter, Feb. 1, 1962.

(Cal. restaurant). *New Yorker*, Apr. 8, 1974.

(Masking flavor). *Fd. Proc.*, July 1976.

(Potentiators). Ibid., Sept. 1976; also *Star Ledger*, Newark, N.J., Aug. 8, 1976.

(Nucleotides). W. O. Caster, "The Nutritionist & His Problems with the New Foods." *Report 3, Inter-Institutional Comm. on Nut.*, in Congressional Hearings on Nutritional Content and Advertising for Dry Breakfast Cereals. Mar. 2, 1972.

Notes

(Egg flavor). *Fd. Proc.*, June 1975.

(Butter flavor). Ibid., Oct. 1975; also *Fd. Prod. Develop.*, Feb. 1974; *Fd. Engineer.*, Dec. 1972; *Snack Fd.*, May 1975; *Fd. Prod. Develop.*, Feb. 1974; *Fd. Proc.*, Feb. 1976; Ibid., Nov. 1975.

(Flavor nuggets). *Fd. Engineer.*, Mar. 1969.

(Vegetable flavors). *Fd. Proc.*, May 1974.

(Agricultural practices affecting fd. flavor). *Flavors & World Fd. Programs*. Internat'l Flavors & Fragrances, 1965; also *Fd. Proc.*, May 1974.

(Potato flavor). *Snack Fd.*, Mar. 1976.

(Mushroom flavor). *Fd. Proc.*, Feb. 1976.

(Green pepper flavor). Ibid., Nov. 1975.

(Onion, garlic flavors). Ibid., Apr. 1975.

(Green soup flavors). Ibid., Feb. 1976.

(Lemonless pie). *Progressive*, Sept. 1975.

(Pineapple flavor). *Fd. Engineer.*, Mar. 1969.

(Fig flavor). *Snack Fd.*, Mar. 1975.

(Raisin flavor). Ibid., Mar. 1974.

(Honeydew melon flavor). *Fd. Proc.*, May 1974.

(Watermelon flavor). Ibid., May 1974.

(Totally syn.). Inglett, *Syn. Fds.*

(Art. orange juice flavor). *Fd. Proc.*, Nov. 1974.

(Natural ice cream). *Am. Dairy Review*, Mar. 1975.

(Caraway seed flavor). *Bakery Prod. & Mktg.*, Jan. 1973. Food processors who resort to imitation and synthetic flavors repeatedly state that a fruit flavor such as strawberry must come from artificial flavors because the supply of such berries would be insufficient to meet consumer demands. This faulty argument needs to be challenged. If demands increase, more farmers would be willing to grow specialty crops such as strawberries, since they could be guaranteed markets for their products at premium prices. Market demands can be met because the laws of supply and demand are flexible. Although strawberries are a highly perishable crop, modern methods of storage, including fast freezing, can hold the crop until it is needed for use in products such as ice cream.

(Artificial & recreated spice particles). *Fd. Proc.*, July 1975; also *Snack Fd.*, Mar. 1974; *Wall St. J.*, 21 Nov. 1977.

(Age of flavor & fragrance ind.). *Chem. Progress*, Sept.–Oct. 1974.

(Jefferson). *Nut. Today*, July–Aug. 1972.

(Scarce vanilla). *Cape Cod* [Mass.] *Times*, Dec. 22, 1975.

(Federal standards). HEW release, July 18, 1960.

(Frozen dessert regulations). FDA information sheet, 1963.

(Demands for vanilla). *Snack Fd.*, Sept. 1975.

("We fooled mother nature"). *Buyers' Guide, Bakery Production*, 1973.

(Maple syrup). *C&EN*, May 22, 1972; also *N.Y. Times*, June 30, 1975; Ibid., May 4, 1975; *Consumer Repts.*, Oct. 1967; Ibid., Oct. 1968.

(Guggenheim). *Media & Consumer*, June 1975.

(Spain—cacao). San Francisco *Chronicle,* May 16, 1976.

(Ghana). *Nat'l. Observer,* Mar. 8, 1975.

(Price—confectioners' coating). *N.Y. Times,* July 22, 1974.

(Candy manufacturers). *Nat'l. Observer,* op. cit.; also L.A. *Times,* Sept. 18, 1974; San Francisco *Chronicle,* Jan. 22, 1974.

(Imitation chocolate flavors, replacers, extenders). *Fd. Proc.,* May 1976; also ibid., June 1975; *Snack Fd.,* June 1975; *Fd. Proc.,* Sept. 1974; ibid., May 1976; ibid., May 1974.

(Baking chips). *N.Y. Times,* op. cit.

(Consumer alerts). *Consumer Protection Newsletter,* Oregon Dept. of Agric., Sept. 1975; also *Consumer Newsletter,* Cal. Dept. of Agric., Jan. 8, 1976.

(Synthetic chocolate ad). *Forbes,* Oct. 15, 1975.

(Poor replicas). Inglett, *Fabricated Fds.,* op. cit.

(Cheap impostors). *New York* mag., Apr. 9, 1975.

(Consumer surveys). *N.Y. Times,* op. cit.

(Aroma). *New York* mag., op. cit.

(French legislation). *Fd. & Cos. Tox.,* Nov. 1969.

(German legislation). M. Pyke, *Synthetic Fds.,* op. cit.; also *Manufacturing Chem. & Aerosol News,* Sept. 1964.

(Lack of uniform laws). *Fd. & Cos. Tox.,* Nov. 1969.

(British rept.). Ibid., Nov.–Dec. 1965.

(Banned U.S. food flavorings). B. T. Hunter, *Mirage of Safety.* N.Y.: Charles Scribner's Sons, 1975.

(Cyclamate status). *FDA Consumer,* June 1976.

(Saccharin status). In autumn 1976 the General Accounting Office recommended that the FDA should either determine that saccharin is safe to be used in food or ban its use. The GAO noted that studies conducted by the health protection branch of the Canadian government on saccharin will not be finished until mid-1978, six years from when the FDA issued its interim regulation. The GAO felt that the FDA's allowance of an interim regulation to remain in effect for so long when safety has not been conclusively established could expose the public to unnecessary risk. The GAO recommended that the FDA should immediately reevaluate the justification for continued use of saccharin in the food supply. *C&EN,* Sept. 27, 1976.

(Canadian studies). *HEW News,* Mar. 9, 1977.

(Gardner & Newell). *New York Times,* 22 Mar. 1977; also see HEW release, 28 Nov. 1977.

(Subacute & chronic toxicity of fd. flavorings & related comps.). *Fd. & Cos. Tox.,* Apr. 1967; also ibid., Sept. 1964; ibid., Aug. 1970.

(FDA's lack of disclosures). *Consumer Bull.,* Oct. 1972.

(Carcinogens). *New York Times,* July 28, 1974.

(Essential oils). *Fd. & Cos. Tox.,* vol. 3, 1965; also ibid., Dec. 1967.

(Mayer). Quoted by D. Zwerling, *Ramparts,* 1971.

Notes

CHAPTER 12. CREATIONS BY THE FOOD COLORISTS

(Pliny). *Food Colors*, Washington, D.C.: Nat'l. Acad. of Sc., 1971.

(Early practices). Ibid.

(Hassall). R. Tannahill, *Fd. in History.*

(Blood-stained fat). *N.Y. Times*, Feb. 7, 1962.

(Sod. sulfite). *Consumer Repts.*, June 1964.

(Sod. nicotinate). *Consumer Bull.*, May 1962; also ibid., July 1963.

(Saithe). *Star Ledger*, Newark, N.J., Nov. 29, 1974.

(Maraschino cherries). Exchange between Vincent A. Kleinfeld, chief counsel to Select Comm. to investigate the use of chems. in fd. products, & John R. Matchet, Ph.D., Bureau of Agric. & Ind. Chem., USDA, Congressional Hearings, *Chemicals in Fd. Products*, 1951.

(Potatoes dyed). J. Kenneth Kirk, FDA, address, Nat'l. Potato Utilization Conference, Gainesville, Fla., May 1, 1961.

(FDA-dyed potatoes). Letter to author, Advisory Opinions Branch, Div. of Industry Advice, Bureau of Ed. & Voluntary Compliance, FDA, Mar. 17, 1964.

(Dyed potatoes not violative). *Fd. Chem. News*, July 16, 1973.

(Dyed sweet potatoes). *Fed. Reg.*, June 21, 1968.

(Winter vs. summer butter). W. A. Price, *Nut. & Physical Degeneration.* Santa Monica, Ca.: Price-Pottenger Nut. Found., 1970.

(Caramel in bread). *Community Nut. Wkly. Rept.*, June 13, 1974. Dry beta carotene beadlets have also been used to impart yellow coloring to bakery products and they materially reduce the cost of the coloring ingredient. The beadlets are made of hydrogenated coconut oil, gelatine, sugar, and modified food starch, with BHA and BHT as antioxidants and methyl propyl parabens, sodium benzoate, and potassium sorbate as preservatives. *Fd. Proc.*, Oct. 1976.

(Pigmenters for poultry). *N.Y. Times*, Nov. 13, 1975.

(Caramel safety studies lacking). *Fd. Colors*, op. cit.

(Danish recommendations). San Francisco *Chronicle*, June 28, 1976.

(Ammonized caramel). *Fd. & Cos. Tox.*, Apr. 1973.

(Gassed tomatoes). *New York Times*, Oct. 19, 1975. According to the United Fresh Fruit & Vegetable Association, "tomatoes picked immature green may redden but will never be good to eat." The latest development in ethylene treatment for tomatoes is an attempt to spray the crops with a chemical that would stimulate ethylene production while the tomatoes are still on the vine. This would cause the crop to turn color at a far faster rate than in nature and all the tomatoes would turn color at the same time. See *The New Yorker*, Jan. 24, 1977.

(FD&C Red No. 2 toxic). *Fed. Reg.* Nov. 16, 1955; also ibid., *Finding of Facts.*

(Hesse). *Fd. Colors*, op. cit.

(Coal-tar dyes). "Food Colors." *FDA fact sheet*, Mar. 1972.

(Azo dyes). *Fd. & Cos. Tox.*, Sept. 1964; also ibid., Aug. 1968. How safe are the U.S. certified dyes in use? The record is far from reassuring. FD&C Blue No. 1, also known as Brilliant Blue FCF, is a triphenylmethane dye which, injected in large

217

doses into experimental rats, produced a significant incidence of malignant tumors. Not permitted in the United Kingdom, the dye is considered "of probable toxicity." It was placed in a category "for which the data available are not wholly sufficient to meet the requirements" of acceptability for food use by the Joint FAO/WHO Expert Committee on Food Additives. The FDA's studies showed this dye produced a high incidence of malignant tumors at the injection site, with some of the tumors metastisizing (spreading, usually through the circulation, and producing tumors in other areas). Despite a high mortality rate, the FDA listed this dye provisionally. Later the FAO/WHO Committee reevaluated it and placed the dye in a category "found acceptable for use in food." In 1969 the FDA gave it permanent acceptance. *Fd. Colors,* op. cit.; also *Fd. & Cos. Tox.,* Sept. 1964; ibid., Aug. 1966; ibid., June 1966. FD&C Blue No. 2, also known as Indigotine or Indigo Carmine, is an indigo-related synthetic dye that was among the seven dyes originally approved. The United Kingdom considered this color provisionally acceptable and the FAO/WHO Committee reported a need "for further information concerning its safety-in-use." The committee requested more data from a two-year study in a nonrodent mammalian species. Short-term studies showed that Indigotine produced blood changes in pigs, and liver abscesses. The FDA's two-year feeding studies with rats and dogs produced malignant tumors at the site of subcutaneous injection. Single oral doses of this dye were absorbed poorly by rats, with little excretion in the urine or bile and most in the feces. The dye's breakdown products were difficult to identify. When this dye was given intravenously to rats, substantial quantities of the dye and relatively smaller amounts of breakdown products were excreted in the bile and urine. These findings of absorption and breakdown products caused surprise and concern. Apart from the dye's food use, it has been used in hospitals to determine kidney function. All injected dye was assumed to be excreted, unchanged, by the kidneys. The presence of breakdown products in the bile and urine can distort results in kidney-function tests. Indigotine is provisionally listed for food use. *Fd. & Cos. Tox.,* Aug. 1976; ibid., Nov. 1969. FD&C Citrus Red No. 2, an azo dye, was fed to rats and mice at different feeding levels for two years. At the highest feeding level, retarded growth and reduced food intake were noted in males of both species. Serious bladder abnormalities were found in postmortem examination. In another experiment this dye, fed to mice up to 80 weeks at various levels, caused increased morbidity and mortality at the three highest feeding levels. An increased death rate for both sexes and an increased incidence of degenerative livers in females were noted. Repeated injections of the dye caused an increased incidence of early malignant tumors in females. As early as 1965, the FAO/WHO Committee classified this dye as one that had been "found to be harmful and that should not be used in food." In 1968, a British study showed that this dye tripled the rate of bladder cancer in rats. In 1969, the FAO/WHO issued a second, stronger statement, warning that since this dye "has been shown to have carcinogenic activity and the toxicological data available were inadequate to allow the determination of a safe limit, the Committee therefore recommends that it should not be used as a food color." Despite this recommendation, the FDA permits this dye to color orange skins not used or intended for further processing. The assumption is that the orange rinds will not be consumed, nor will the dye migrate into the fruit's edible

Notes

portion. But the assumption is faulty. People use orange skins in marmalades, as candied rinds, powdered rinds, with roast duck, and in cocktails, and children sometimes peel oranges with their teeth. *Fd. & Cos. Tox.*, Aug. 1966; also ibid., Nov. 1966; *Fd. Colors*, op. cit.; *Br. J. Cancer*, vol. 22, 1968; *New Zealand Med. J.*, vol. 73, 1971. FD&C Green No. 3, also known as Fast Green FCF, is a triphenylmethane dye. It produced a significant incidence of sarcomas at the site of repeated subcutaneous injections in rats. The FDA's chronic toxicity studies with rats, dogs, and mice showed the dye "without significant toxic effect" at the dosage levels fed, and the agency listed the dye provisionally. *Fd. Colors*, op. cit.; also *Fd. & Cos. Tox.*, Aug. 1966. Orange B, an azo dye, is a relatively new food dye, with research development begun in 1953. In 1966, the FDA approved its use in food but restricted it to color the casings or surfaces of frankfurters and sausages at specified levels. Lifetime feeding studies of Orange B, with three different animal species, failed to substantiate the dye's safety. Liver nodules formed in dogs whose diets contained 2 percent or more of the dye. The significance of this nodule formation is unknown. The dye has not yet been evaluated by the FAO/WHO Committee, either for classification or to establish an acceptable human daily intake level. *J. Pharm. Exp. Ther.*, vol. 26, 1956; also *Fd. Chem. News*, June 11, 1973; *C&EN*, Oct. 17, 1966. FD&C Red No. 3, also known as Erythrosine, is a xanthene dye, and was among the seven originally approved for food use. Although the FAO/WHO Committee found this color "acceptable for use in foods," the dye was given only temporary acceptable status. The committee recommended further studies concerning the dye's metabolism with several species of animals, and preferably in humans. Additional studies were requested, on the mechanism responsible for the dye's effect on iodine levels bound to blood plasma. The German Research Institute for Food Chemistry in Munich found that Erythrosine had a slight but significant mutagenic (gene altering) effect on a common microorganism, *Escherichia coli*. In testing other xanthene dyes, the same institute found that they displayed a similar mutagenic effect on bacteria. These findings are noteworthy since frequently, when substances are mutagenic they are also carcinogenic. FD&C Red No. 3 was provisionally approved by the FDA. *Fd. & Cos. Tox.*, July 1964; also ibid., Dec. 1972; ibid., Apr. 1973. FD&C Red No. 40, also known as Allura Red AC, is an azo dye. In 1971, when the FDA gave this dye permanent approval for food and drug use, it was hailed by the industry as "the only important new food color certified by FDA since 1939." In 1974, the FDA gave similar approval for this dye in cosmetics. The dye was developed to replace the banned FD&C Red No. 4. Once again, in 1976, when FD&C Red No. 2 was banned, the industry turned to FD&C Red No. 40 as an alternative. In addition to the FDA's granting permanent status, this dye was approved in other countries. However, the FAO/WHO Committee requested additional studies, and the Canadian government, which had examined the same data as the FDA, along with updatings, refused to approve the dye. The Canadians charged that the "evidence submitted by the manufacturer with respect to the safety of the product was inadequate." The Canadians were concerned especially about the long-term rat feeding tests designed primarily to determine if the dye caused cancer. The tests were originally designed for twenty-four months, but pulmonary disease ravaged the animals and the tests were terminated.

219

The Canadians questioned the adequacy of the tests, doubting that a twenty-one-month test was sufficient to detect a long-term effect such as cancer. The only significant effect in these abbreviated tests was that the dye, fed at the highest level, caused moderate growth suppression during the first year. Although the FDA recommends full lifetime studies, conducted with two rodent species, the agency had given permanent approval to the dye on the basis of studies with a single rodent species, the rat. The FDA, the Canadian officials, and international food and health authorities requested further tests to provide additional data. Newer studies, termed by the FDA as "highly preliminary," showed that in week forty-one of a seventy-eight-week study a disproportionately high percentage of mice developed "premature and unexpected malignant lymphomas" (cancer of the lymph glands). Follow-up studies were inconclusive. FDA officials admitted that the dilemma with FD&C Red No. 40 bore a striking resemblance to the banned FD&C Red No. 2; namely, that the safety of the dye had neither been proved nor disproved. Officials face similar dilemmas regarding cyclamate and saccharin. *Science*, Feb. 27, 1976; also *N.Y. Times*, Feb. 28, 1976; HEW *News*, Feb. 28, 1976; *Wall St. J.*, Mar. 1, 1967. With the banning of several red dyes, new products have been introduced, using solids of dried juice from concord and red grapes, cranberry, and red cherry. *Fd. Proc.*, Oct. 1976. FD&C Yellow No. 5, also known as Tartrazine, is a pyrazolone dye. It is one of the few dyes that has been universally accepted in countries where food coloring is permitted. In 1969 the FDA granted permanent listing for this dye. Nevertheless, numerous cases of human sensitivity to Tartrazine have been pinpointed, with a wide range of adverse reactions reported especially by allergists. *Ann. of Allergy*, vol. 33, 1974; also *Fd. & Cos. Tox.*, June 1968; Ibid., Aug. 1973; *Skin & Allergy News*, July 1975; *Med. Tribune*, June 9, 1971; Ibid., Aug. 18, 1971; *Rev. Allergy*, vol. 24, 1970; B. F. Feingold, *Introd. to Cl. Allergy*, Springfield, Illinois: Charles C. Thomas, 1973; L. D. Dickey, ed., *Cl. Ecology*, Springfield, Illinois: Charles C. Thomas, 1976. FD&C Yellow No. 6, also known as Sunset Yellow FCF, is an azo dye. Although extensively studied for its effects on animals, information is lacking regarding the dye's metabolism. In short-term studies with miniature pigs and rats, as well as in acute toxicity studies with rats and mice, animals suffered from slight diarrhea, although no adverse effects were noted on growth or food consumption. In a seven-year study, dogs fed 2 percent of the dye in their diet suffered eye defects, sometimes leading to blindness. *J. Pharm. Exp. Ther.*, vol. 136, 1962; also *Fd. & Cos. Tox.*, Oct. 1967.

(Mattson). Washington *Post*, 16 June 1977.

(Dyed fd.). *Br. Med. J.* Apr. 22, 1961.

(Bicknell). Ibid., May 6, 1961.

(Lederberg). *Vogue*, July 1976.

CHAPTER 13. NATURAL VERSUS SYNTHETIC VITAMINS

(l-amino acids). M. G. Wohl & R. S. Goodhart, *Modern Nut. in Hlth. & Disease.* Phila., Pa.: Lea & Febiger, 1968; also *Dairy Council Digest*, Nov.–Dec. 1972. Few

Notes

d-amino acids are present in natural metabolism; frequently, only in lower organisms or pathological tissues. On the average, d-forms are two to four times more slowly absorbed than l-forms. The rate of absorption differs for each amino acid, with methionine the easiest and histidine the slowest. With more highly purified and concentrated diets—that is, with the use of synthetic vitamins—the dl-form of tryptophan is less adequate than that of an equal weight of l-tryptophan, and the difference is accentuated at lower intakes of the whole diet. The effect of pure l- and of pure d-tryptophan on the growth of rats showed that the d-form is distinctly less suitable for growth purposes than the l-form. Under certain conditions it is possible that the body splits the dl-form, in which case it can make use of the l-form but may have difficulty with the d-form. Pure methionine can be utilized in promoting the growth of test animals, provided that five other essential amino acids (histidine, leucine, lysine, phenylalanine, tryptophan) are present in the natural l-form. When these amino acids are present in the dl-form, d-methionine is inferior. The same is observed with the d-isomer of phenylalanine. d-valine is ineffective when leucine is supplied in the synthetic d-5 form but can be utilized when natural l-leucine is present. d-valine is completely inert nutritionally. Another important aspect of the use of the d-isomers is the relative failure of the kidney to retain these, contrasted with the corresponding l-form, so that the nutritional effectiveness of a particular d-isomer may be as much dependent on renal clearance as on the body's ability to utilize it for growth and maintenance. Thus, the body first encounters problems in absorbing and utilizing synthetic amino acids, and then tries to get rid of these unnatural substances as fast as possible. The problems of absorption and utilization are encountered in the intestines. Transport of the l-isomer is achieved by means of enzymatic reactions, whereas that of the d-isomer is achieved by some process of diffusion. Thus, the natural forms of amino acids are actively and selectively absorbed by the body; the unnatural ones force their way into the body by osmosis. The use of synthetic amino acids may lead to unexpected complications. The adequacy and safety of diets using such materials are unpredictable.

(dl-amino acids to fortify fds. & supplement animal feed). *C&EN*, Dec. 23, 1974; also ibid., Dec. 24, 1973; USDA press release, July 29, 1974.

(GRAS). *Fed. Reg.*, July 26, 1973, part 2. For a technical description of the structural differences of amino acids derived from natural protein and synthetic amino acids, see *Science*, Mar. 1, 1974.

(Pantothenate). R. J. Williams, *Nut. in a Nutshell*, New York: Dolphin, 1962.

(Pure vs. concomitants). Not only is it necessary to have all components of vitamin complexes present, with their known and unknown nutrients, but they must be intricately interrelated with other vitamin complexes, as well as minerals and other nutrients, in order to function optimally. For this reason, *single* vitamin deficiencies are rare.

(Overstimulation of glandular system). Letter to author; E. E. Pfeiffer, M.D.

(Close relationship). I. Jennings, *Vits. in Endocrine Metabolism.* Bioflavonoids strengthen the body's tiniest blood vessels, the capillaries. Bioflavonoids were discovered by Dr. Albert Szent-Gyorgi when he was attempting to isolate ascorbic acid in

Hungary. Being uncertain about the nature of the bioflavonoids, he tentatively called them Vitamin P. He used that letter "because it was toward the end of the ABCs, still far from the letters used in vitaminology. In the event that I was wrong, the name could be dropped without causing any difficulty. I also chose 'P' because the names of most pleasant things, in Hungarian, begin with the letter P." Szent-Gyorgi, *The Living State*, New York: Academic Press, 1972. The medical literature contains numerous references to the importance of bioflavonoids in human health. See especially "Bioflavonoids & the Capillary Conf. Rept." *Ann. N.Y. Acad. of Sc.*, vol. 61, 1955; M. S. Biskind & W. C. Martin, "The Use of Citrus Bioflavonoids in Respiratory Infections. *Am. J. of Digest. Diseases*, vol. 21, 1954; *Med. Wrld. News*, July 28, 1975; R. C. Robbins, *Internat'l J. of Vit. & Nut. Res.*, Apr. 1973. Despite the medical reports demonstrating therapeutic roles of bioflavonoids in a wide range of health problems, the FDA continues to hold the opinion that "no nutritional claims may be made for ingredients such as bioflavonoids . . . having no scientifically recognized nutritional value." "Vitamins, Minerals & FDA," *FDA reprint*, 1974.

(Crude natural extracts released slowly in body). *Bull. of Nut'al. Med.*, Mar. 1973.

(Naturally derived vits. more effective). *J. of Immunol.*, Aug. 1942; also *Scandinavian Veterinary*, vol. 30, 1940; *J. of Nut.*, vol. 21, 1941; *Nature*, Jan. 1, 1952; E. W. H. Cruickshank, *Fd. & Nut., the Physiological Bases of Human Nut.* Edinburgh: E. & S. Livingstone, 1946; H. R. Rosenberg, *Chem. & Physiology of the Vits.* N.Y.: Interscience, 1945; *J. of Nut.*, vol. 28, 1944; *Nut.*, vol. 7, Summer 1953.

(Natural Vitamin E). Information supplied both by Eastman Chem. Prods. & Gen'l Mills, suppliers of Vitamin E products.

(Free radicals). *C&EN*, June 7, 1971. Prof. H. von Euler, Nobel Prize winner for chemistry, wrote that Vitamin C action of the ascorbic acid found in fruits and vegetables is more effective than the synthetic ascorbic acid. W. Schuphan, *Nut'al. Values in Crops & Plants*. London: Faber & Faber, 1965.

(Syn. ascorbic acid more effective with foods). *Lancet*, Dec. 11, 1937. In one experiment, Canadian researchers found that synthetic ascorbic acid was as effective as the natural form. Moreover, "the higher serum levels and lower excretion of ascorbic acid that occurred with synthetic ascorbic acid compared with orange juice indicate slightly greater bioavailability of the synthetic vitamins. The synthetic ascorbic acid provided in freshly opened fruit drinks seems to be well utilized." What was *not* indicated in this statement was that the small amount of real fruit juice (up to 10 percent) contained in the fruit juice drink may have provided the necessary accessory factors. *J. of Am. Diet. Assoc.*, vol. 64, 1974. Some artificial fruit drinks, with synthetic ascorbic acid added, are touted as being products with more Vitamin C than real fruit juices. The implication that such products are nutritionally superior is patently false. Comparative studies with synthetic ascorbic acid and frozen orange juice concentrate showed better utilization of the vitamin when it was within its natural group of substances. This is particularly true with patients who lack hydrochloric acid in their gastric juices. *Münschener Mediz. Wochenschrift*, vol. 103, no. 20, May 1961.

(German World War II prisoners). Ibid., vol. 76, 1951.

(Norwegians). *Hippokrates*, vol. 5, 1955.

Notes

(Later experiments). *Nut. Notes*, Oct. 1963. A natural source of ascorbic acid was shown to be superior to a synthetic one in resistance to infection. The ascorbic acid level in the blood could be raised to a specific amount by giving far smaller quantities of orange juice than ascorbic acid alone. The Vitamin C content of leukocytes, the white blood cells that fight infection, is considered as a very sensitive indicator of serum and tissue saturation of ascorbic acid. *Münschener Mediz. Wochenschrift*, vol. 120, 1961. Dentists have reported that they were better able to control oral bleeding, tenderness, redness, and swelling of tissues in the oral cavity of patients when bioflavonoids (accessory factors present in Vitamin C) were given in conjunction with synthetic ascorbic acid. E. Cheraskin, et al., "Diet & Nut. in Oral Hlth. & Disease," paper, conference, Univ. of Ala. Med. Center, Birmingham, Ala., Mar. 1965. Allergists report that a synthetically derived vitamin may cause a reaction in a chemically susceptible person, when the same type of vitamin, derived from natural substances, may be well tolerated. Clinical reactions are frequent with ascorbic acid, and Vitamin B_1 (thiamine). T. G. Randolph, *Human Ecology & Susceptibility to the Chem. Environment*. Chicago, Illinois: Charles C. Thomas, 1962; also E. W. Kailin, testimony, congressional hearings, *Proposed FDA Supplement Regulations*, Wash., D.C.: Mar. 6, 1970; S.D. Lockey, "Sensitivity to FD&C Dyes in Drugs, Fds. & Bev.," in L. D. Dickey, ed., *Clinical Ecology*. Springfield, Illinois: Charles C. Thomas, 1976. Many of these human experiences have been confirmed in animal studies. Like human beings, guinea pigs need to obtain their supply of ascorbic acid from food sources. For this reason, these animals frequently are used to test different forms of ascorbic acid and study the importance of its accessory factors, the bioflavonoids (Vitamin P). As early as 1937, and again in 1948, guinea pigs were shown to have better tissue storage of ascorbic acid from foods than of ascorbic acid in crystalline form. *J. of Nut.*, vol. 14, 1937; *Nature*, Apr. 10, 1948. In 1953 investigators reported that guinea pigs were more protected from scurvy and hemorrhaging by the use of ascorbic acid from natural sources than the synthetic form. In recent years, guinea pigs continue to be used in studies comparing the effectiveness of natural versus synthetic ascorbic acid. The synergistic action of ascorbic acid was shown to be increased about 50 percent by the presence of Vitamin P, found in all citrus fruit. *Biokhimiya*, vol. 19, 1953. In another study, scurvy was induced in guinea pigs by feeding them a diet that lacked ascorbic acid. Then one group was given 1-ascorbic acid from black currant juice concentrate, another group an equivalent amount from dried acerola juice, and a third group synthetic ascorbic acid in water. The overall growth rate was greatest with the group fed black currant juice, intermediate with acerola juice, and lowest with synthetic ascorbic acid. *J. of Sc. of Fd. & Agric.*, vol. 22, 1971. In another study, both the aggregation of blood cells and the capillary resistance were measured in guinea pigs. The measurement of aggregation of blood cells is used as an indicator for scurvy: the higher the aggregation, the more severe the condition. First the animals were deprived of ascorbic acid in their diet. Then the acid was restored to their diets in differing amounts or with the addition of one of the bioflavonoids (rutin, hesperidin, or naringen). The animals receiving ascorbic acid plus the bioflavonoids showed significantly less aggregation of blood cells and higher capillary resistance than those receiving as-

223

corbic acid alone. All three bioflavonoids were equally effective, but no treatment restored capillary resistance completely nor did any of the treatments completely abolish aggregation of blood cells in these animals that had all suffered early deprivation of ascorbic acid. *Internat'l. Zeitschrift fur Vitaminforschung,* vol. 36, 1966. At times, other animal species have been used to test natural versus synthetic ascorbic acid, and the results have shown differences. Although ascorbic acid was found to play an important role in enhancing monkeys' resistance to poliomyelitis infection, optimal results were obtained from natural sources of ascorbic acid and less favorable ones with the synthetic form. *J. of Exp. Med.,* vol. 70, 1939. Scurvy in a litter of puppies was not cured by synthetic ascorbic acid, whereas fresh orange juice resulted in what was described as a "spectacular recovery." Synthetic ascorbic acid, given to rabbits, failed to raise the ascorbic acid level in the blood, which was raised by supplying the animals with fresh cabbage leaves. *Veterinary Record,* vol. 74, 1962.

(Vitamin A complex). When Vitamin A was synthesized, it was learned that more than one form exists in natural products. Although the variations in the different structures are not very great, some possess higher activity than others. The different variations of molecular structure are shown diagramatically by Pyke in *Synthetic Fd.,* op. cit. He draws his work from *Vitamins & Hormones,* vol. 18, 1960.

(Superiority of natural Vitamin B complex). M. Wohl, ed., *Dietherapy,* 1945.

(Treating malnourished infants). Sternmann, *Schweizer Mediz. Wochenschrift.* Lucerne, 1941.

(Animals on syn. Vitamin B failed to thrive). *Scandinavian Veterinarian,* vol. 30, 1940.

(Yeast factor). *J. of Nut.,* Apr. 10, 1958.

(Pyridoxine). Jennings, *Vits. in Endocrine Metabolism,* op. cit. Also, from 1968 to 1971, new metabolites of Vitamin D_3 were isolated, identified, and synthesized. In almost every respect they are ten to fifteen times more active than Vitamin D_3 in preventing and curing rickets in rats and chicks. *Nut. News,* Dec. 1973.

(Structural differences of Vitamin D_2 and D_3). Jennings, *Vits. in Endocrine Metabolism,* op. cit.

(Dangers of excessive Vitamin D). *Consumers' Res. Mag.,* Oct. 1930.

(Chick exp.). *J. of Biol. Chem.,* vol. 97, 1932.

(Hypercalcemia in Gr. Brit.). *N.Y. State J. of Med.,* Apr. 1, 1966.

(Toxic effects of Vitamin D_2). *JAMA,* Feb. 24, 1953.

(FDA proposed limit of Vitamin D products). HEW news release, Dec. 13, 1972.

(Dairies using Vitamin D_3). Letters to author.

(Vitamin C, adjusted with ascorbic acid). *J. of Nut. Ed.,* vol. 4, 1972; also *JAMA,* July 2, 1973; ibid., Aug. 27, 1973; ibid., Nov. 5, 1973.

(Vitamin E potencies). Eastman Chem. Prods. & Gen'l. Mills, op. cit.; also Remmington, *The Vitamins,* vol. 11; *Biochem.,* vol. 2, 1963; *Nut. Reviews,* vol. 25, 1967; *J. of Animal Sc.,* vol. 27, 1968.

Notes

CHAPTER 14. MOTHER'S BREASTMILK VERSUS
INFANT FEEDING FORMULAS

(Jelliffe). Statement, congressional hearings, *Maternal, Fetal & Infant Nut.*, June 5,
1973.
(Percent of nursing mothers). A. Berg, *The Nut. Factor.* Wash., D.C.: Brookings
Institution, 1973.
(8 wks. nursing). *Cutis,* July 1974.
(Promoting formulas with professionals). *N.Y. Times,* Sept. 11, 1975; ibid., June 9,
1976. A documentary TV film, "The Unfinished Child," shown on a major network in
1976, was sponsored by a formula maker. Although one scene showed breastfeeding,
several scenes showed formula feeding. Moreover, the still photograph that served as
the film's logo was of a mother bottlefeeding her child. The sole organization to
promote breastfeeding is La Leche League International, a voluntary women's group
with headquarters at Franklin Park, Illinois.

The failure to use human milk represents a senseless waste of a national resource.
Breastmilk, an existing high-protein infant food, is being replaced by other protein-
rich foods, usually based on cows' milk. In these times of critical assessment of our
food policy and nutritional needs, this questionable practice needs reexamination.

In Third World countries the promotion of infant feeding formulas raises even more
critical issues. The stigma against breastfeeding, similar to the earlier stigma in the
developed countries, began to spread to Third World countries after World War II.

The UN's Protein Advisory Group warned that early abandonment of breastfeeding
in poor families in underdeveloped countries can be disastrous to infants. A pediatrician warned that "in underdeveloped countries, breastfeeding stands as a bulwark between infant mortality and survival." Some women in Third World countries may lack
the skills or resources to use infant feeding formulas properly. They may not understand the printed instructions accompanying the formulas. In order to make the expensive product last longer, they may dilute it excessively and thus lower its nutritive
value. Or the water available for mixing the formula may be contaminated. The scarcity and expense of fuel may make bottle sterilization impossible. These factors combine and create problems. Diarrhea and other illnesses are far more common among
such formula-fed infants than breastfed ones. The mortality rate of formula-fed infants
in Third World countries is far higher than those breastfed exclusively.

As breastfeeding has declined in the last two decades in Third World countries,
severe forms of malnutrition have appeared in infants at an early age. Frequently,
malnutrition is now manifest at eight months of age. This trend is especially alarming,
in view of recent knowledge about the critical importance of early nutrition for proper
physical and mental development.

Although official agencies and governmental authorities have urged a return to
breastfeeding in Third World countries, commercial counterpressures are stronger.
Infant feeding formula advertising is extensive and aggressive. Ethically, some techniques are questionable. A leading critic, Dr. Jelliffe, accused these commercial companies of fostering "commerciogenic malnutrition."

Pressures to have mothers accept infant feeding formulas in Third World countries take subtle forms. "Milk nurses" or "mothercraft workers" may be hired by formula manufacturers to visit new mothers in the maternity wards and later at homes. At times these workers are merely detail persons pushing their products. Often they are dressed in traditional nurses' uniforms, which lend a spurious authoritative air to their positions. Some are paid on a commission or bonus basis. According to an official of the Protein Advisory Group, mothers exposed to such pressures "will accept anything from someone dressed in a white coat."

The companies involved in such practices do not deny them. "It's a highly competitive market, and [the] hospital is an obvious place to reach the new mother." The companies claim that they are simply "fulfilling a need."

These tactics have been deplored by many critics, including nutritionists, officials, public interest groups, and church organizations. All are concerned about excessive promotion of products by manufacturers, and the resulting discouragement of breastfeeding. Indirectly, abandonment of breastfeeding in Third World countries leads to increased malnutrition, disease, and mortality. For an in-depth study of this problem, see T. Greiner, "The Promotion of Bottlefeeding by Multinational Corporations: How Advertising & the Health Professionals Have Contributed," *Cornell Internat'l. Nut. Monograph Series*, no. 2. Ithaca, N.Y.: Cornell Univ., 1975; ibid., T. Greiner, "Strategies for Solving the Bottle Feeding Problem," no. 4, 1977; also *Science*, May 9, 1975.

(Cost of formulas for Third World). *Science*, Apr. 5, 1974, op. cit.

(Health problems). *N.Y. Times*, Apr. 6, 1976. Although dilution of formulas to make them last longer is a problem in Third World countries, the opposite problem exists in the affluent West. An overconcentrated formula with low water content and a high concentration of solutes cannot be handled by the functionally immature infant kidney. *Mod. Med.*, Jan. 1, 1976.

(Earlier malnutrition manifest). *N.Y. Times*, Apr. 6, 1976, op. cit.

(Differences in basic composition of milks). *Am. J. of Cl. Nut.*, Aug. 1971.

(6 polysaccharides). *Compt. Rend. Acad. Sc.*, 263D, 1966.

(Antiinfective properties). *Science*, May 9, 1975. Also, antibodies in breastmilk have been found helpful in the treatment of celiac disease in infants. *Lancet*, Apr. 3, 1976.

(Jelliffe). *Am. J. of Cl. Nut.*, Aug. 1971, op. cit.

(Protein). S. J. Fomon, *Comparative Nutritive Values of Human Milk & Various Milk Preparations*, Iowa City, Iowa: Dept. of Ped., Univ. of Iowa, 1973. Infant feeding formulas, as being inferior to breastmilk, are recognized in the Dietary Food Regulations as formulated by the FDA in 1973. Such formulas must have a minimum 1.8 Protein Efficiency Ratio (PER) value. If less, the label of the formula must state, "This product should not be used as the sole source of protein in the infant diet."

The large size of curds in cows' milk was the incriminating factor in intestinal obstruction found in newly born infants who had been bottlefed with whole cows' milk or milk powder from cows' milk used in formulas. The babies were from five to fourteen days old. It was necessary to operate on most of them in order to eliminate the obstructing curds.

Notes

(Nucleotides). *Am. J. of Cl. Nut.*, Aug. 1971, op. cit.

(Substitution of oil). Fomon, op. cit., points out that the reasons for the difference in absorption between human milk fat and cows' butterfat are not completely understood. The differences probably relate more to the arrangement of fatty acids on the triglyceride molecule than to a difference in content of individual fatty acids. Palmitic acid, which accounts for approximately one fourth of the weight of total fatty acids in each of these fats, is nearly equally distributed in butterfat among the three positions of the triglyceride, but palmitic acid in human milk fat is primarily esterified at the 2-position. Because of the mode of action of pancreatic lipase and because the 2-monoglyceride of palmitic acid is well absorbed, whereas free palmitic acid is poorly absorbed, the difference in location of palmitic acid on the triglyceride is likely to account for much of the difference in digestibility of the two fats. Pancreatic lipase specifically hydrolyzes the fatty acids esterified in the 1- and 3-positions of a triglyceride, with no hydrolysis of the majority of fatty acids that are esterified in the 2-position. Thus, the feeding of human milk will result in the presence in the intestinal tract of most of the palmitic acid in the form of the readily absorbed 2-monopalmitin, whereas feeding of butterfat will give rise to a much greater quantity of the poorly absorbed free palmitic acid.

(Cholesterol levels). *Mod. Med.*, June 15, 1975.

(Moderate level of cholesterol desirable). Fomon, op. cit.

(Human brain development). Among the substances in breastmilk that are vital for brain formation are cerebrosides, a group of white, waxlike basic glycolipids found especially in the brain and other nerve tissue. On hydrolysis it yields a fatty acid, sphingosine, and a sugar, galactose. *Ecologist*, Jan. 1975.

(Lipase). *Am. J. of Cl. Nut.*, Aug. 1971, op. cit.

(Linoleic acid). Ibid.

(Vitamin A content). M. Hardy, op. cit.

(Folic acid). J. L. Mount, *The Fd. & Hlth. of Western Man.*

(Ascorbic acid). Ibid.

(Vitamin D). Ibid.

(Vitamin E). Hardy, op. cit.; also M. Ebon, *The Truth About Vitamin E.* New York: Bantam, 1972. The ratio of Vitamin E to PUFA is discussed in Anemia, Information Sheet 24, rev. 1972. Franklin Park, Illinois: La Leche League Internat'l.

(Lactose). Hardy, op. cit.

(Trace minerals). M. Hambridge & D. O'Brien, *Trace Metals in Childhood Nut.*, Ross Labs, 1973.

(Sodium). Hardy, op. cit.

(Edema). *Med. Wrld. News*, Feb. 5, 1965.

(Iron). If a mother has been extremely anemic during her pregnancy, her infant is likely to be born with a low store of iron. Hemoglobin iron concentrated at birth varies greatly. The level may determine the iron requirements of the infant for the following months. Transferrin, a special substance identified in human but not in cows' milk, is responsible for a system of transporting iron from the mother to the infant. At the time of delivery, if the umbilical cord is allowed to drain its valuable

227

iron-rich blood before the cord is clamped the infant will receive more iron. Anemia, Information Sheet 24, op. cit. Ivan Kochan, professor of microbiology, Miami University, Oxford, Ohio, reported that iron-fortified foods may enhance the ability of bacteria in the body to cause disease. Breastmilk contains iron-binding proteins that make iron unavailable to bacteria, whereas cows' milk contains very little iron-binding protein. *Wall St. J.*, Apr. 8, 1976.

(Transferrin). *Nut. Today*, Summer 1969.

(Anemia). *Blood*, vol. 13, 1958.

(Aggravation of iron-deficiency anemia). *J. of Peds.*, vol. 60, no. 5.

(Produce vs. cereals). *Nut. Today*, Summer 1969, op. cit.

(Hemoglobin levels). Anemia, Info. Sheet 24, op. cit.

(Vitamin E-iron). *New Engl. J. of Med.*, Nov. 28, 1968.

(Ascorbic acid). H. F. Meyer, *Infant Fds. & Feeding Practices.* Springfield, Ill: Charles C. Thomas, 1960.

(Copper). *Fed. Proc. Part 11, Trans. Suppl.*, vol. 24, July–Aug. 1965.

(Repts. of copper deficiency). Hambridge & O'Brien, *Trace Metals in Childhood Nut.*, op. cit.

(Zinc : copper ratio). *Am. J. of Cl. Nut.*, July 1975; also *C&EN*, Apr. 14, 1975.

(Convulsions). *JAMA*, vol. 154, 1954; also *Nut. Reviews*, Oct. 1955. Although only about fifty infants developed convulsive seizures, some four hundred infants suffered from central nervous system disorders ranging from mild to grave.

(Formula suit). San Francisco *Examiner*, Feb. 9, 1975.

(Skin eruptions). *Br. Med. J.*, vol. 1, 1961; also ibid., vol. 2, 1962; *Arch. of Diseases of Children*, vol. 37, 1963.

(Vitamin E shortage in feeding formulas). *Council of Better Bus. Bureaus*, Dec. 26, 1973.

(Recommendation for Vitamin E fortification). *Am. J. of Cl. Nut.*, Mar. 1967.

(Stress reactions). Jennings, *Vitamins in Endocrine Metabolism*, op. cit.

(Long-storage loss of nutrients). *Better Nut.*, Oct. 1971. Surveys were conducted by the Consumer Federation of America in typical states: Arizona, California, Illinois, Louisiana, and Oregon. Spot checks showed that 57 percent of all infant feeding formulas selected randomly were on shelves in stores beyond the eighteen-month period. The state of Virginia had passed a law requiring pull dates on infant feeding formulas. Nevertheless, a survey in that state in May 1974 showed that 20 percent of grocery and 37 percent of drugstores still had outdated infant feeding formulas for sale.

(Colostrum). *Canadian Consumer*, Aug. 1975.

(White cells in colostrum). *Med. Wrld. News*, June 16, 1975. Lactoferrin, a protein found abundantly in colostrum, appears to be involved in the killing of bacteria by phagocytes. Also see *Br. Med. J.*, vol. 11, 1972; *Med. Wrld. News*, Nov. 19, 1971.

(Preemies). *Med. Wrld. News*, Sept. 27, 1974.

(Flora). *Nut. News*, Dec. 1975.

(Gastroenteritis). *Nut. Reviews*, May 1965.

(Morbidity & mortality). *Am. J. of Hyg.*, vol. 2, 1922; also *JAMA*, Sept. 8, 1934; ibid., June 11, 1935; *Peds.*, Dec. 1974.

Notes

(Liverpool). *Med. Wrld. News,* Sept. 13, 1974.

(Otitis media). *Emergency Med.,* Dec. 1971.

(Dental decay). *The Express,* Easton, Pa., Oct. 30, 1975. Dr. Julius Ozick, a dental faculty member at New York University, suggested before the annual meeting of the ADA that "bottle baby caries" (cavities) may result from bottlefeeding too frequently or too long. Also see *German Tribune,* Apr. 25, 1976.

(Nursing bottle syndrome). *Nut. News,* Feb. 1975; also *Star Ledger,* Newark, N.J. Mar. 21, 1974. The American Society of Dentistry for Children launched a campaign to warn parents of the harmful effects to teeth by allowing extended bottlefeeding.

(Obesity). Chicago *Tribune,* Feb. 7, 1976.

(Multicell obesity). Montreal *Star,* Sept. 13, 1972; also *Med. Wrld. News,* Sept. 7, 1973. In addition to the problems of obesity started in the first year of life, Dr. A. W. Myres, Nutrition Bureau, Department of Health and Welfare, Ottawa, attributed other health hazards to infant feeding formulas. Hypernatremic dehydration has also become a common clinical disorder. An important factor in its etiology is high calorie/solute feedings. Studies in the British Commonwealth have shown that infants fed infant feeding formulas plus commercial infant foods in the early months of life, among other things, have an abnormally high level of blood urea. Myres reported that these problems are unlikely to arise in the completely breastfed infant because of the low solute content of human milk and because the infant regulates its own intake. Of further interest, suggested Myres, are the results of animal studies indicating that patterns of enzyme synthesis may be altered by infant formula feeding and early weaning. These findings serve to emphasize how relatively little we know about the long-term consequences of early nutritional influence, wrote Myres, and they suggest that the modern trend away from breastfeeding needs to be reevaluated. *Nut. Today,* May–June 1974.

(Behaviorist psychologists). *Sweeteners, Issues & Uncertainties. Academy Forum.* Wash., D.C.: Nat'l. Acad. of Sc., 1975. The problem of forcing the infant to finish everything in the bottle was raised by Professor Alfred Harper, chairman of the Department of Nutritional Sciences, University of Wisconsin. Elsewhere, Dr. René Dubos suggested that malnutrition can take many different forms; it may even be caused by excessive infant bottlefeeding. Little is known about the physical and mental effects of a nutritional regimen that differs in quality from that of the mother's milk. Infants fed an abundant diet tend to become large eaters as adults. Dubos suggested that such acquired dietary habits may, in the long run, have physiological drawbacks and "it would not be surprising if they did not have behavioral manifestations. Rapid growth and large size may not be unmixed blessings." René Dubos, *So Human an Animal.* New York: Charles Scribner's Sons, 1968.

(Paté de foie babies). Jelliffe, testimony, congressional hearings, *Maternal, Fetal & Infant Nut.,* op. cit.

(Bonding). *Maternal & Child Hlth,* Mar. 1972.

(Thromboembolism). *Med. Tribune,* Jan. 15, 1968. Emboli was high in nonlactating women in a two-year study of nearly a thousand women in Cardiff, Wales.

(Maternal anemia). Congressional hearings, *Maternal, Fetal & Infant Nut.,* op. cit.

(Delays pregnancy). *Consumer Repts.,* Mar. 1977; *Nut. Today,* Jan.–Feb. 1977;

"*Why Nurse Your Baby?*," pamphlet, Franklin Park, Ill: La Leche League Internat'l, undated.
(Experimental & uncertain). Jelliffe, Montreal *Star*, Sept. 13, 1972, op. cit.

CHAPTER 15. FDA FAVORS AND ENCOURAGES IMITATION
FOOD MANUFACTURE

(Early definition of imitation). Press release, USDA, Feb. 9, 1962; also *Fd. Chem. News*, Mar. 26, 1973.
(Recommendations for changed definition). White House Conf. on Fd., Nut. & Hlth., final report, Wash., D.C.: GPO, 1970; also *Fd. & Drug Pkging*, May 22, 1975; *FDA Consumer*, Oct. 1975.
(McNamara). *Fd. & Drug Pkging*, op. cit.
(Fd. technologists). *N.Y. Times*, Jan. 30, 1975.
(Fed. of Homemakers). *Newsletter*, Fed. of Homemakers, Spring 1975.
(Dyson). *N.Y. Times*, Dec. 12, 1975; also N.Y. *Post*, Dec. 11, 1975.
(New appeal upheld FDA). *Am. Dairy Review*, May 1975; also *Fd. Chem. News*, June 14, 1976; Ibid., July 16, 1973.
(Imitation mayonnaise). *Chr. Sc. Monitor*, Sept. 3, 1975; also *Fd. Proc.*, May 1975.
(Jacobson). *Chr. Sc. Monitor*, op. cit.
(FDA nutritional labeling). HEW *News*, Jan. 17, 1973.
(Merchandising gimmick). *Bus. Wk.*, Feb. 3, 1973.
(Marketing scoop). Ibid., Feb. 24, 1973.
(Nutrient variability). *Fd. Proc.*, July 1974.
(Variations in milk). *Am. Dairy Review*, Dec. 1974.
(Variations in meat). *Current Status of Nutrition Labeling*, Meat Inst. Res. Conf. Ames, Iowa: June 1972.
(United Fresh Fruit & Veg. Growers suit). *Am. Fruit Grower*, May 1973; also *Nut. Notes*, June 1973; Ibid., Sept. 1973.
(400-fold difference in iron). *Monthly Supply Letter*, United Fresh Fruit & Veg. Assoc., Dec. 1973.
(Babcock). *Fd. Chem. News*, Feb. 12, 1973.
(Briggs). *Fd. & Nut. News*, Apr. 1973.

CHAPTER 16. THE NEW LOOK IN ANIMAL FEED

(Immunologic protection in colostrum). F. W. R. Brambell, "The Transmission of Passive Immunity from Mother to Young," in A. Neuberger & E. L. Tatem, eds., *Frontiers of Biology*. Amsterdam: North-Holland Pub., 1970.
(Gastrointestinal tract permeable to protein). *Biochem. Bio-phys.*, Acta 181, 1969; also *Res. Vet. Sc.*, vol. 14, 1973.
(Colibacillosis). *J. Comp. Path.*, vol. 80, 1970.

Notes

(Replacers—diseases). *Br. J. Nut.*, vol. 14, 1960; also ibid., vol. 16, 1962; ibid., vol. 17, 1963.

(Immunoglobulins denatured). *Peds.*, Dec. 1974.

(Intraperitoneal gammaglobulin substitution). *Canadian J. Animal Sc.*, vol. 41, 1961; also ibid., vol. 4, 1964; Ibid., vol. 35, 1972.

(NPN replacers). Circular S-230, Gainesville, Florida: Fla. Agric. Exp. Sta., Univ. of Fla., Nov. 1974.

(Urea). *Dairy Council Digest*, Jan.–Feb. 1970.

(Deficiency of urea). *Good Fd.*, Sept. 1973.

(Dairy cows). *Agric. Res.*, June 1966.

(Cement kiln dust). USDA press release, 14 Dec. 1977.

(Feathers as replacer). *Service*, Aug. 1968.

(Polyethylene plastic pellets). *Fd. Chem. News*, June 23, 1969.

(Wood fibers). *C&EN*, Aug. 5, 1974.

(Pulp residue). Ibid., Sept. 14, 1974.

(Hemicellulose). News release, Am. Chem. Soc., Sept. 12, 1968.

(Aspen-fed steers). *Fd. Management*, Mar. 1976.

(Wood chips). News release, USPHS, June 1968.

(Sawdust). *South Dakota Farm & Home Res.*, Spring 1972.

(Sawdust-alfalfa). Ibid., Summer 1973.

(Rumose). *Am. Agric.*, Oct. 1975.

(Ground newspapers). *Agric. Res.*, Feb. 1971.

(Contaminants in newsprint). *C&EN*, Dec. 18, 1972; also *Environment*, Jan.–Feb. 1973.

(Wholesome agric. wastes). *C&EN*, Oct. 5, 1974.

(Sewage–algae). Ibid., May 22, 1972.

(Sludge). Ibid., Sept. 23, 1974.

(Municipal garbage). *Fd. Engineer.*, Nov. 1973.

(Pelleting animal waste). Press release, USDA, Apr. 21, 1975.

(DPW more profitable in animal feed than as fertilizer). *Poultry Times*, Southeastern ed., July 31, 1974.

(DPW-cattle, sheep). Press release, USDA, July 29, 1974.

(DPW-protein levels). *Poultry Times*, July 31, 1974, op. cit.

(USDA exps.). *Meat Research, An Agric. Res. Serv. Progress Rept. Info. Bull. 375.* Wash., D.C.: USDA, Jan. 1975.

(DPW-laying hens). *Natural Farmer Newsletter*, Putney, Vt.: Dec. 1971–Jan. 1972.

(DPW-economics). *Sc. News*, May 11, 1974; also *Chem. News*, May 13, 1974.

(High-protein bacteria). *C&EN*, Apr. 10, 1972.

(Image problem). *N.Y. Times*, Sept. 8, 1973.

(Potential contaminants in DPW). *Poultry Times*, Apr. 14, 1975.

(DPW more valuable for fertilizers & energy sources). Blair T. Bower, *Alternatives for Handling Chicken Manure*. Arlington, Virginia: by author, July 1975.

(DPW-state approvals). *N.Y. Times*, May 5, 1975. In 1973, the Ceres Land Company, one of the nation's large beef cattle feeding operations, began to include in the normal diet of its herd substantial portions of feed derived from cattle manure. By the end of December 1975 DPW had been approved in at least a half dozen states. In Georgia its use was restricted to dried cage-layer droppings free of drug residues. This is a restriction that will be difficult to monitor. *Farm J.*, Dec. 1975.

(1976 relaxed regulations by FDA). *Fd. Chem. News*, Feb. 16, 1976.

(Waste as oyster feed). Information #71, New Engl. Marine Resources Info. Program, Apr. 1975.

(Substitute fds. for lobster). Ibid., Information #14, July 1975.

(Substitute fds. for rainbow trout). *Arch. of Path.*, May 1961.

(Liver cancer in trout). W. C. Hueper & W. D. Conway, *Chemical Carcinogenesis & Cancers*. Chicago, Illinois: Charles C. Thomas, 1964.

(Trout mortalities). *N.Y. Times*, July 18, 1961.

(SCP: oil-based). *Chem. Wk.*, Aug. 8, 1973.

(Yeast: petroleum based). *Fd. Proc.*, July 1974.

(Japanese SCP). *C&EN*, Mar. 5, 1973; also *Foreign Agric.*, Sept. 1973; *Mother Jones*, Aug. 1977.

(Italian SCP). *N.Y. Times*, May 30, 1976. A large fermenter plant in England is producing SCP for animal feed using methanol as a substrate. *C&EN*, Oct. 11, 1976.

(*Lancet*). *Lancet*, Nov. 11, 1972.

(Synthetic amino acids). Press release, USDA, July 29, 1974.

(Animals on totally syn. diet). *Agric. Res.*, Apr. 1965.

(1975 livestock management conference). *N.Y. Times*, May 5, 1975.

Chapter 17. REAL FOOD: AN ENDANGERED SPECIES

(Davis's predictions). *J. Roy. Soc. Arts.*, vol. 114, 1966.

(Pitfalls of amino acid fortification). *Arch. Biochem. & Biophys.*, vol. 58, 1955. Amino acids differ in their configurations. They are essentially small simple-moleculed structures. Manufacturing them involves combining an enormous complexity of subtle polymers that constitute protein. Methionine was the first amino acid synthesized commercially and is the only amino acid of which the two optically active forms, *levo* and *dextro* isomers, are equally well used in human and animal nutrition. The chemical synthesis of lysine yields a mixture of two lysine molecules, one in which the structure is of the biologically unavailable *dextro* configuration. In synthesizing threonine it was difficult to separate the desirable isomer from the mixture of isomers often produced simultaneously during synthesis. The unwanted dl-allothreonine form is produced more readily than the desirable dl-threonine form.

(Protenoid). Pyke, *Man & Fd.*

(Cellulosic waste). *C&EN*, Apr. 12, 1976.

(Canteen automation). Pyke, *Automation: Its Purpose & Future.*

(Institutional automation). *Inst./Vol. Fding.*, Feb. 1976.

Notes

(Unmanned supermarkets). *U.S. News & World Rept.*, Nov. 3, 1975; also *N.Y. Times*, Aug. 2, 1975.

(NASA). *Fd. Management*, June 1976; also *N.Y. Times*, Apr. 8, 1976.

(Fd. tech. not nutritionists). A. E. Bender, "The Nutritionist in Industry," in J. Yudkin & J. M. McKenzie, eds., *Changing Fd. Habits*. London: MacGibbon & Kee, 1964.

(Syn. fd. & global hunger). *C&EN*, ibid., Jan. 27, 1975; Aug. 4, 1975; ibid., Sept. 1, 1975.

(Syn. fds. no solution). *Sc. Am.*, Sept. 1976.

(Foreign ventures). J. Keats, *What Ever Happened to Mom's Apple Pie?* Boston: Houghton Mifflin, 1976.

(Snack fds.). *Snack Fd.*, Sept. 1976.

(Inadequacies of semi- and all-syn. diets). E. W. H. Cruickshank, *Fd. & Nut., the Physiological Bases of Human Nut.*

(Public view of sc. & tech.). *Science*, Apr. 11, 1975.

(AAAS study). *C&EN*, May 26, 1975.

(Tech. phenomenon). J. Ellul, *The Technological Soc.*, N.Y.: Knopf, 1964.

(Illich). I. Illich, *Medical Nemesis, the Expropriation of Hlth.* Toronto: McClelland & Stewart, 1975.

INDEX

Accum, Frederick, 89
Albrecht, William, 133–34
Alexander, J. C., 187
Allied Chemical Corporation, 171
American Academy for the Advancement of Science, 193
American Academy of Pediatrics, 142
American Butter Institute, 63
American Medical Association, 142
amino acids: forms of, 135–36; fortification with, 232; synthesis of, 188–89; utilization of, 220–21
Amoco Food Company, 181
analogs: cheese, 61; meat, 8, 21, 23, 26–30
Anderson, Robert, 43
animal feed: current practices in, 170–83; rational program of, 183–85; replacers in, 173–83; roughage in, 175–77; traditional, 172
artificial foods, definition of, 8
ascorbic acid, 134, 136–39, 222–24
automation and food, 189–92
Awake (juice concentrate), 86

Babcock, M. J., 168
bacon, simulated, 24
Baron von Redberry (cereal), 80
base of food supply narrowed, 6–7
Battista, O. A., 92
"beef grill steak," 15
beef "steak," 15
beef strip, 15
3-4 benzpyrene, 182
Berglas, Alexander, 132
Bicknell, Franklin: on food colors, 121, 131; on hydrogenated fat, 68–69
bioflavonoids, 137, 221–24
Borden, Inc., 41
Boyland, E., 120
breastfeeding, 146–60

Briggs, George M.: on fabricated food, 11–12, 163; on nutritional labeling, 168
Bureau of Chemistry, 130
Burros, Marian, 93–94
butter: and food color, 126; synthetic flavor of, 105–7; versus margarine, 63–70
butterine, 65–66
Butz, Earl L., 31

calamus oil, 116–17
California Milk Advisory Board, 42
Cameron, Allan G., 41
caramel, and bakery products, 127–28, 215
carbohydrates, synthetic, 91–97
Carlson, Anton J., on food, 187; on vitamins, 133
Carnation Company, 40–41
Carper, Jean, 81
Carroll, Kenneth, 6
Carver, George Washington, 23
Caster, W. O., 105
cellulose: in bread, 93–95; in feed, 176–77
cement kiln dust in animal feed, 175
Center for Science in the Public Interest: on caramel, 127–28; on labeling, 166
Chapman, R. A., 122
cheese: flavorings, 59–61; imitation, 54–62
chemical additives: in imitation fruit drinks, 87–88; industrial sources of, 4; in nondairy creamers, 46–47; in nondairy whipped toppings, 51; in synthetic butter flavor, 107
Cherry, Rona, 55
cherry, synthetic, 81
Chevreul, Eugène, 65
chicken, simulated, 24, 27
chicken feathers as flour replacer, 97

child feeding and soy protein, 30–32
chocolate, synthetic, 113–15
cholesterol issue, 23, 33–39, 44, 152, 206, 211, 227
cis isomer, 67–70, 211–12
citrus beverages, 7
Claiborne, Craig, 21
Clausi, A. S., 80
Coca Cola, 193
coconut oil, 6
coffee creamers, 43–50
coffee lighteners. *See* coffee creamers
colorants: and adulteration, 123–30; in butter, 126; to mask blemishes, 124–25; in synthetic food, 120–31; toxicity of, 217–20
composition of milk, 150–56
computerization and food, 189–92
Conklin, Gordon, 171
convenience foods, 8
cooked beef-fat tissue solids, 18
cortex, 113
Coumarin, 116
creamers, nondairy, 43–50
Cruickshank, E. W. H., 63
Crum, George, 74

Davis, J. G., 188
dehydrated food, 73–74
Denenberg, Herbert, 162–63
Desmond, Ruth: on analogs, 165; on imitation food labeling, 162
d-forms, 135, 144, 220–21, 232
diabetics: and egg substitute, 38; and textured vegetable protein, 25
dietary fiber, 93–95
dl-forms, 135, 144, 220–21, 232
doublethink, 162–63
Downey, William, 2
DPW. *See* dried poultry waste
Drawert, F., 99
dried poultry waste (DPW), 178–80
Dubos, René, 186
Dulcin, 116–17

Dunbar, Paul B., 188
Dyes: Azo, 130–31, 217–19; from coal tar, 4, 130, 217–20; FD&C Citrus Red No. 2, 129; FD&C Red No. 32, 129
Dyson, John S., 165

EFA. *See* essential fatty acids
Efron, Marshall, 2
egg: restructured, 13–14; substitute, 33–39; yolk and food dye, 127
Egg-Beaters®, 33–35, 206–7
Ellul, Jacques, 194
Elrick and Lavidge, Inc., 11
encapsulation, 107, 110
"engineered" food: definition of, 8; as meat substitute, 20; and world hunger, 192–93
Enstrom, Elmer W., 187
essential fatty acids, 68–69
essential oils for food flavoring, 119
ethylene gassing of tomatoes, 128, 217
extender, meat, 24
extruded food: future of, 10, 71; machinery to make, 4; onion rings, 74–75

fabricated food: definition of, 8; development of, 5–7
fast food service, 7
fats in milk, 151–52
Federal Food Standards, 165
Federal Trade Commission: and fruit drinks, 85; and margarine, 63; and natural foods, 161
Federation of Homemakers, 162, 164
feed, animal, 170–85
feeding formulas: deficiencies in infant, 156–57; history of, 146–60; and obesity, 229; and Third World, 225–26
fiber, dietary, 93–95
filled milk, 48–50
fish, restructured, 13, 18
Fite, George L., 146
flaked food, 73

Index

flavor: enhancers, 104–5; intensifiers, 104–5; potentiators, 104–5
flavorants: lack of safety in, 116–18; as replacers of food, 105–8; supply of, 215
flavoring seed replacers, 109
flavors, synthetic, 99–119, 215
Florida Citrus Commission, 84–86
Fomon, Samuel J., 152, 226–27
food: and automation, 189–92; engineering, 192–93; and FDA, 194; and molecular rearrangement, 12; standards, 8; and toxicology tests, 118–19; and USDA, 194; and vending, 189–92
food color: and egg yolk, 127; and oranges, 129–30; and poultry skin, 126–27; provisional status of, 131; and toxicology, 130–31, 217–20
Food and Drug Administration: and approval of DPW, 179–80; and consumer protection, 194; and dietary fiber, 94–95; and food color, 125–31; and imitation labeling, 60–62, 161–69; and orange juice standards, 86–87; and toxicity tests, 118–19; and vitamins, 132, 134, 142
Food Research and Action Center, 31
Food Standards Committee's Report on Flavoring Agents, 116
Forbes, Allan, 94
formulated food, 8
fortification of plant protein, 28–29
Fox, Brian A., 41
French Academy of Medicine, 116
french fries, restructured, 75
Friedlander, Paul J. C., 40
Fritzsche Dodge & Olcott, Inc., 2
fruit: drinks, 86–89; flavors, 102, 108–9; restructured, 71–78; synthetic, 81–82, 102, 108–9

Galbraith, John Kenneth, 79
gamma valerolactone, 119
garbage in animal feed, 178
General Foods Corporation, 84–86

General Mills, Inc., 80
Giedion, Siegfried, 186–87
Gilbert & Sullivan, 42
Golana, 61
granulated food, 73
Gresham's Law, 113
Grimm, Charles, 2
Guggenheimer, Elinor, 113

Hale, William C., 3
ham "steak," restructured, 15
Handing, Earl M., 121
Hardin, Clifford, 31
Harland, Barbara, 94–95
Hassall, Arthur Hill, 123
Herber, Lewis, 172
Hess, Jerry, 193
Hesse, Bernhard C., 130
Hightower, Jim, 40–41
Hoelzel, Frederick, 91–92
Holmes, Oliver Wendell, 148
human milk, composition of, 150–56
hydrogenation, 67–70, 209–10

Illich, Ivan, 195
imitation: cheese, 54–62; colorants, 120–31; FDA's definition of, 60–62; food, 8; fruit drinks, 86–89; labeling of, 161–69; rice, 91; sour cream, 52
"inconvenience" foods, 9
industrial machinery applied to food processing, 4
infant feeding formulas, 146–60, 225–27, 229
Institute of Food Technologists, 131
International Flavor and Fragrances, Inc., 2
iron-deficiency anemia, 155–56, 227–28

Jacobson, Michael, 166
Jelliffe, Derrick B.: on decline in breastfeeding, 148; on values of human milk, 146–47; on Third World, 225
Johnson, Ogden C., 132

Index

Oberleas, Donald, 94
O'Connor, Harvey, 89–90
oil of calamus, 116–17
oleomargarine, 65
onion rings, extruded, 74–75
orange drink, synthetic, 83–87, 213
Orange Plus, 85–86
oranges and food colorant, 129–30
Orwell, George, 162
oxalic acid, 28, 205

pancake syrup, 112–13
pantothenate, 136
Patterson, Robert, 100
pelletized food, 72
Peng, A. C., 21
pennyroyal, 116
PepsiCo, 193
PER. See protein efficiency ratio
Perkins, Sir William Henry, 4
Pet Food Company, 40
petroleum as food substrate, 181, 204
phytic acid, 28, 205
plant protein: and essential amino acids, 27–28; and fortification, 28–29
Plenora, 36
Pliny, 122
polarized light, 135, 138
pollulan, 96–97
pork, restructured: breakfast strip, 15; "choplet," 15; simulated link, 24; simulated patty, 24
potato chips: definition of, 75–77; restructured, 75–78
Potato Chip Institute International, 76
potato skin: food colorant with, 125–26
poultry: and food colorant, 126–27; restructured, 14–15
protein: efficiency ratio (PER), 22; in human milk, 151; plant, 22; quality of, 27; structure of, 10; synthetic flavors of, 102–3
Protein-Calorie Advisory Group, United Nations, 149, 225

puff-exploded food, 72–73
purine nucleotides, 105
Pyke, Magnus: on egg, 34–35; on egg substitute, 36; on food synthesis, 1; on inconvenience food, 9; on margarine, 63; on processing food, 11; on protein, 10; on synthetic color, 120–21; on synthetic flavor, 100; on synthetic produce, 82
pyrazolone dyes, 220
pyridoxine: feeding formulas deficient in, 156; synthesized, 140

quassia wood, 116

raisins, synthetic, 82
Ralston Purina Company, 31
Randhawa, M. S., 79
RDAs. See Recommended Daily Allowances
Recommended Daily Allowances (RDAs), 166
refining of food, 4
replacer ingredients: in animal feed, 173–83; in fish feed, 180–81; in flavoring seeds, 109; in flour, 89–98; in spices, 109–10; use of, 5–6; vanillin as an example of early, 110–12
restructured food: beef "steak," 15; beef strip, 15; definition of, 8; egg, 13–14; fish, 13; fish "fillet," 18; french fries, 75; fruit, 71–78; ham "steak," 15; meat, 14–19; meat and textured vegetable protein, 15–16; pork breakfast strip, 15; pork "choplet," 15; potato chip, 75–78; poultry, 14–15; techniques of manufacturing, 12–19; veal "steak," 15; vegetables, 71–78
rice, imitation, 91
Root, Waverley: on meat flavor, 170; on public taste, 80–81
roughage replacers in feed, 175–77
rumose as roughage replacer, 176–77

239

Index

"unfood," 194
Unilever Corporation, and margarine, 70
United Fresh Fruit and Vegetable Association, 167–68
U.S. certified food colors, 130
U.S. Department of Agriculture: and food, 194; and synthetic animal feed, 182–83
U.S. Pharmacopeia, 137

vanilla, 110–12
Vanillamark, 111
vanillin, 110–12
Vaupel, Susanne, 31
veal "steak," restructured, 15
vegamine, 140
vegetables: restructured, 71–78; synthetic, 81–83; synthetic flavors of, 102–3, 107–8
vending, food, 189–92
Verrett, Jacqueline, 119
Vitamins: A, 139; B, 139–40; C, 134, 136–39, 222–24; D, 140–42; E, 138, 143–45, 157

Vitamins: in milk, 153; natural versus synthetic, 132–45
von Hoffman, Nicholas: on food processing, 11; on food supply, 3
von Liebig, Justus, 4

waffle syrup, 112–13
Warner Jenkinson Company, 120
Weiking Eiweiss, 35
Weininger, Jean, 163
whipped topping: saturated fat in nondairy, 7; history of nondairy, 50–52
White House Conference on Food, Nutrition, and Health, 164
Wisconsin Alumni Research Foundation, 60
Wodicka, Virgil O., 162
wood pulp in bread, 93–95
Woodbury, Robert M., 147
Wynder, Ernest, 6

xanthan gum, 96
xanthene food dyes, 130–31, 219–20